COMPARING URBAN SERVICE DELIVERY SYSTEMS

Structure and Performance

Volume 12, URBAN AFFAIRS ANNUAL REVIEWS

COMPARING URBAN SERVICE DELIVERY SYSTEMS

Structure and Performance

Edited by
VINCENT OSTROM
and
FRANCES PENNELL BISH

Volume 12, URBAN AFFAIRS ANNUAL REVIEWS

 **SAGE PUBLICATIONS** / BEVERLY HILLS / LONDON

For information address:

SAGE PUBLICATIONS, INC.
275 South Beverly Drive
Beverly Hills, California 90212

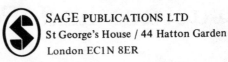

SAGE PUBLICATIONS LTD
St George's House / 44 Hatton Garden
London EC1N 8ER

Printed in the United States of America

International Standard Book Number 0-8039-0470-3 (cloth)
                                  0-8039-0680-3 (paper)

Library of Congress Catalog Card No. 76-6312

FIRST PRINTING

CONTENTS

Introduction

FRANCES PENNELL BISH

☐ A PERENNIAL ISSUE IN HUMAN AFFAIRS is whether the structures or forms of human association make any difference when measured by values that are important to human life. Efforts to look at political experience writ large have often led to the conclusion that political structure makes relatively little difference. Parliamentary institutions, for example, do not seem to work when transplanted to different social and cultural settings. Also the gap between expectations and performance in the structuring of socialist societies is substantial. As a result of the perceived failure of political institutions to "work as they should," many scholars have turned to cultural, social, or economic variables as the critical factors affecting the quality of human life. The structure or forms of human institutions are assumed to have little effect.

The contention that structure makes little difference has been powerfully reinforced by contemporary scholars who have sought to identify determinants of public policies. Using expenditures as an indicator of output, these studies have sought to identify variables which explain or account for variations in expenditures among different nations, states, or provinces and local units of government within nations. The recurrent finding has been that structural variables—such as party competition, differences in electoral systems, and executive organization—seem to make little difference. The more important variables in explaining variations in levels of public expenditure appear to be wealth, urbanization, industrialization, and educational achievement.

Few social scientists would dispute the contention that these variables have a profound effect upon conditions of modern life. But as the British parliamentarian Shirley Williams points out, these findings are "not particularly germane for the politician." She goes on to observe:

> One can hardly go out to increase the urban proportion of the population in one's state simply to increase the level of spending on education. Likewise with research plotting, for instance, the relationship between the existence of liberal democracy and literacy rates in the world. Such findings to the politician are, at the most, suggestive.

What would be "germane for the politician" are studies which address those variables over which policy makers have some measure of control.

The principal instrument or tool available to policy makers is the reallocation of authority by assigning differential capabilities and limitations for the decisions that individuals make in their relationships with others. A basic mechanism of social control is, thus, the restructuring of the rules of the game so as to create a different structure of incentives and encumbrances that will affect the choices people make.

In spite of a strong tradition in contemporary political science alleging that "structure makes little difference," the idea that structures do make a difference is reflected in the many studies and proposals advocating reorganization of service delivery systems as a major solution to public policy problems. These recommendations for reform reflect the implicit assumption that a restructuring of the rules of the game will affect the choices people make and, thereby, the performance of public service delivery systems.

The logic underlying these proposals is described by Vincent Ostrom in the first article in this collection. Ostrom suggests that one of the most promising developments in the study of human and political experience is the shift from power and influence to service delivery systems as the focus of theoretical and empirical analysis. Studies of service delivery systems focus upon the relationship between structure and performance in the production and consumption of essential public services. Analysis of the relationship between structure and performance requires reference to the *calculations* and *criteria* to be used to inform choices about problems of *design, implementation,* and *evaluation.* Central to these is the question of what conception or theory is to inform inquiry and action.

The traditional approach to political inquiry begins with the state as a monopoly over the legitimate use of force in a society and with the presumption that all states are essentially unitary and hierarchically organized in nature. It was on this basis that Max Weber formulated his "ideal type" model of bureaucratic administration. This ideal type underlies many of the recommendations for reform and reorganization of public service delivery systems. Within this context, increased professionalization and consolidation of multiple governmental units into highly integrated and bureaucratically organized units of relatively large scale have generally been viewed as essential prerequisites for improving the performance of public service delivery systems.

An alternative basis for conceptualizing the organization of public service delivery systems derives from the theory of public goods, from the works of such scholars as Madison, Hamilton, and de Tocqueville and the efforts of modern public choice economists. This alternative conception begins with individuals as the basic units of analysis and uses the structure of incentives and payoffs inherent in different decision rules and different types of goods and services to derive conclusions about the human consequences likely to follow when service delivery systems are organized in different ways. Scholars using this approach have questioned the assumption that highly integrated and bureaucratically organized forms of governance will *necessarily* lead to long-run improvements in the quality of human life. These scholars have suggested that complex and overlapping systems of governance, operating under appropriately constituted decision rules, may lead to better results—particularly within the context of the highly complex and changing patterns of interaction characteristic of modern urban and industrialized societies.

The social scientist (and the politician) is, thus, confronted with two radically different theoretical bases or concepts for organizing society for the improvement of human welfare. The tasks to be pursued in the development of a social science that will permit the making of choices from among the array of structural possibilities are, as Ostrom suggests, first, to learn to use language as a tool for reasoning through the implications that follow from specifiable assumptions and postulates and, second, to undertake empirical research designed to test hypotheses derived from specifiable assumptions and postulates. Such inquiry should be carried out in a variety of different cultural and social settings and at a variety of different levels of analysis.

Variety both in cultural and social settings and in levels of analysis is evident in the essays included in this volume. However, all share a common concern not only for the development of a "warrantable body of knowledge" relating structure to performance but for the use of language to reason through the logical implications that follow from specifiable assumptions and postulates.

Beyond this common concern for the development of theory and the undertaking of research on the relationship between structure and performance, a number of themes are developed in this volume. The first of these is "complexity"—both in the forms of urban governance and in the environments in which these governments operate. This theme is particularly stressed in Part II, which includes articles by Daniel J. Elazar, Kenneth Hanf, and Philip Sabetti. In these articles the emphasis is also upon the cultural, social, and political factors that may affect both the structure and the performance of local public service delivery systems.

Elazar's account of the organizational arrangements used for producing local public services in Israel suggests a structural arrangement of considerable complexity. Indeed, in Israel the very word for "organization of polity" or "government" *(leharkiv)* means to compound. The same word is used to describe complexity.

The production of local public services in Israel is compounded in the sense that it involves the activities of and interactions among a variety of different public and private institutions—including those that express the unique relationship between the state of Israel and the Jewish people, those of the various confessional minorities living in Israel, those established by the various ideological parties, and those of the governmental, private, and cooperative sectors. The government sector is itself compounded with a relatively large number of cities, federations of cities, and special districts.

In his description of evolving organizational arrangements in the German Democratic Republic, Kenneth Hanf also notes a structure of growing complexity. He suggests that most analysts have tended to stress the integrated nature of the bureaucratic systems through which essential public services are delivered in socialist countries to the exclusion of the "significant pieces of political action" that may occur at the local level and in the relationships among and between levels of government and the economic sector. Such analyses, he argues, overestimate the effectiveness of central direction and control and the efficacy of hierarchical chains of command.

A striking feature of the East German system, as described by Hanf, is the system's considerable complexity involving production activities on the part of an array of public and quasi-public institutions. A variety of coordinating mechanisms have developed among these producers of public services—including the rather extensive use of contracting between and among local assemblies and economic enterprises, and the development of joint authorities and associations of communities.

In contrast to the extremely complex sets of organizational characteristics of East Germany and Israel, arrangements used for producing local public services in Italy have traditionally followed a form more closely allied with the Weberian "ideal type" of a highly integrated and bureaucratically organized structure of governance. Contemporary political arrangements in Italy were fashioned with the intent of creating a nation-state with a single center of authority, an exclusive monopoly on the legitimate use of force and a single overarching system of public administration. Sabetti argues that the choice of a monocentric state on the part of the architects of modern Italy—both in the 1860s and post World War II—represented an explicit attempt to forge diverse communities of people into a single community of interest. Arrangements for producing and delivering local public services in Italy—essentially a decentralized, but hierarchically organized state system with minimal autonomy at the communal level—reflect this desire. Some communes have attempted to overcome problems associated with urban public service delivery through the establishment of neighborhood governments. The central government has also recently initiated reforms resulting in the creation of regional governments. Whether these innovations will be successful remains to be seen; they, in any case, reflect a movement toward a somewhat more complex system of governance.

The delivery of local public services in Israel and East Germany and increasingly in Italy is thus characterized by "complexity" —particularly with respect to the interactions that occur among and between levels of government and other segments of the society.

In contrast to the growing complexity in *organizational arrange-*

ments described in these three articles, the articles by Roger W. Benjamin, Arthur B. Gunlicks, and Ruud J. Vader in the third part of the volume deal with efforts to simplify the organization of public service delivery systems through consolidation and increased professionalization.

Benjamin describes the changes that have taken place in England as a result of the 1972 Local Government Act. This act resulted in a reduction in the total number of local authorities from over 1,000 to a little over 400. The act was based largely on the assessments of the British Royal Commission which found the following weaknesses in English local government: fragmentation of governmental units, division of responsibility between upper and lower tiers within counties, the small size of many authorities, an arbitrary division between town and country, and an unsatisfactory relationship between local authorities and the public.

Benjamin critiques the reasoning used in support of the Local Government Act on a number of grounds—arguing, among other things, that developments in post-industrial societies require modifications in the structure of local political institutions that will make them more responsive to rapid change, equity considerations, and citizen needs and expectations. He questions whether simplifying the structure of governance will achieve these ends.

Gunlicks describes and critiques the reasoning underlying local government reforms in West Germany, which, since the mid-sixties, have resulted in a reduction in the number of cities and towns from 24,500 to fewer than 11,000 and a reduction in the number of counties from 425 to 250.

The constitutional argument for reform in West Germany rests on the principles of "equality" and "freedom" which are seen as complementary. Within this context, it is argued that uniform or equal provision of services can only be achieved if all local governmental units are strong enough financially to provide a high level of services. Dependence upon the national government for financing or service delivery would inevitably result in a weakening of local government—a circumstance that would threaten the viability of local governments as democratic institutions. The argument has also been made that an industrial welfare state requires a system of local government quite different from that in the past—particularly with the change in emphasis from law and order to the provision of services. Other assessments of the structural weaknesses inherent in West Germany's system of local government included overlapping and fragmented jurisdictions, the need for coordination of effort and unity of purpose, the need for highly qualified and professional administrators, and the small size of the most local governmental units.

Reforms proposed for the organization of police services in the Netherlands are part of a larger reform effort involving the entire administrative system. Proposed changes in the organization of police services call for the elimination of all municipal police departments. This function would be assumed by a reorganized national state police, which currently serves rural areas and small municipalities.

Like the reform efforts in England, proposed reforms in the Netherlands are based, in part, on the assumption that municipalities have been unable to adapt to changes brought about by urbanization and industrialization and are too small to provide an effective and highly professionalized police force.

In the case of Israel, East Germany, and Italy, then, we find structural arrangements moving away from a model of the Weberian type to one of increasing complexity. In the case of reforms proposed or undertaken in England, the Netherlands, and West Germany, structural arrangements appear to be evolving toward a model more in accord with Weber's ideal type. Although several of the authors (Gunlicks and Sabetti, for example) provide some evidence of the impact of these reforms on local government performance, all suggest that more rigorous assessments of the consequences need to be undertaken.

Assessment of the consequences will require not only a further specification of the hypotheses and assumptions implicit in reform proposals but methodologies for examining and isolating their impact. The articles by Werner W. Pommerehne and Bruno S. Frey, Bengt Owe Birgersson, and Lars Stromberg in the last part of the volume are examples of explicit attempts to test hypotheses regarding relationships between structure and performance and to develop methodologies appropriate to this task.

Pommerehne and Frey compare the performance of public and private refuse collection agencies in 103 Swiss cities. Noting first that there are a priori reasons to expect problems or difficulties with respect to both public and private production, they go on to examine the performance of the two types of producers—using "efficiency" as their evaluative criterion. They find that, at least in the cities under consideration, private production appears to be associated with lower per unit costs. In concluding, they make the important point that private production can retain its advantage over public production only if competitive pressure is maintained—pressure which may be a function of the structure of incentives established through governmental efforts.

The Birgersson and Stromberg articles use data from studies in 36 Swedish communes to examine the relationship between structure and responsiveness to local citizen demands and expectations. Birgersson examines two fundamental questions: first, do larger communes offer a higher level of services than smaller communes and, second, are citizens living in communes with higher service levels more satisfied than those living in communes with low service levels? He finds that larger communes generally offer a higher level of services—where the level of services is measured in "quantitative" rather than "qualitative" terms. He also finds, however, that citizens living in communes with higher service levels are more frequently dissatisfed with the services provided than those living in communes with lower service levels. The association of higher levels of service with lower levels of citizen satisfaction creates a "service paradox."

The service paradox may result from a "whetting of the appetite" on the part of citizens for higher and higher levels of public services. Or, it may result from the failure of larger, more urbanized communes to successfully assess and meet citizen preferences with respect to local public service delivery. The latter explanation is at least partially supported by the finding that participation rates tend to be lower in larger communes. Larger communes also tend to have a more highly professionalized public service, which, if there is substantial disparity between professional and citizen assessments of what is wanted and needed, could very well result in a service paradox where service levels are indeed high but do not meet citizen preferences.

This possibility is reinforced by Stromberg's article which examines the relationships between and among the type of commune, socioeconomic representativeness, and opinion representativeness. Citizen and councillor attitudes toward expenditure levels and service priorities are used to measure opinion representativeness, and data on the socioeconomic characteristics of citizens and councillors are used to measure socioeconomic representativeness. Stromberg finds little relationship between socioeconomic representativeness and opinion representativeness and fairly good agreement between councillors and citizens with respect to priorities as among types of local governmental services. He also finds, however, that councillors, in the case of all parties and all types of communes, differ systematically from citizens in their willingness to increase or expand communal expenditures—a phenomenon which, he suggests, may be accounted for by differences in the roles or positions held by the two groups. Differences both in opinion representativeness and socioeconomic representativeness are greatest in the larger, more urbanized communes. The findings that councillors differ systematically from citizens in their willingness to spend money and that opinion representativeness and socioeconomic representativeness are lowest in the larger communes may contribute to Birgersson's service paradox.

The collection of essays included in this volume suggest that the structures or forms of human association do make a difference when measured by values that are important to human life. Structures or forms of human association may not always be the most important determinants of policy decisions or individual actions. Many variables affect human behavior and differences in the structures or forms of human association represent only some of these variables. Choices are, however, possible from an array of alternatives in structuring decision-making arrangements. If such choices do make a difference, it is important that those who choose be informed by warrantable knowledge that can stand the basic evidentiary test of whether a given form of action will yield predictable consequences. We need a social science that addresses itself to the relationship between structure and performance.

In concluding, we wish to thank our colleagues at the Workshop in Political Theory and Policy Analysis at Indiana University, whose efforts contributed substantially to the volume. Roger B. Parks, Steven Percy, Eric Scott, and Elinor Ostrom read and commented on a number of the papers. An extremely able secretarial staff including Marsha Brown, Andrea LaPeyre, Gill Nevin, and Mary Zielinski did an excellent job of typing and retyping manuscripts and assembling the volume for publication.

Part I

THEORY

1

Structure and Performance

VINCENT OSTROM

□ SOME NEW ORIENTATIONS and concepts are being developed which promise major advances in the social sciences. One of these is a shift from studies of power and influence to studies of service delivery systems. We might hypothesize that the structure of decision-making arrangements will affect the way that power and influence are mobilized and exercised. Service delivery indicates the effect that exercise of power and influence has upon the potential welfare of citizens as recipients of services. The potential effect of structural conditions upon performance is revealed by the quality of service that those who exercise power and influence render as a function of their positions in society.

Another recent conceptual development is the explicit attention being given to a theory of public goods as distinguished from private goods and services. Public goods theory has stimulated a major reconsideration of the types of nonmarket decision-making arrangements that are appropriate to the organization of *public economies* as distinguished from *market economies*. This reconsideration of

AUTHOR'S NOTE: *This chapter draws, in part, upon a paper, "Political Orders and Public Service Structures," presented at the International Institute of Management Conference on Interorganizational Decision-Making and Public Policy, Berlin, June 18-20, 1975.*

nonmarket decision structures is evolving into a theory that has major implications for generating research hypotheses and assessing the policy implications that follow from reliance upon different structural arrangements. A knowledge of different nonmarket decision-making arrangements and their effect on the delivery of public goods and services may also enable people to design public service delivery systems that better serve their purposes when measured in relation to specifiable performance criteria.

These intellectual developments will be considered, in turn, in the sections of this essay that follow: (1) power, influence, and service, (2) the nature of public goods, (3) the organization of public service delivery systems, and (4) coping with alternatives, implications, and choices.

POWER, INFLUENCE, AND SERVICE

POWER AND INFLUENCE

Power has been a perennial preoccupation in the study of government and politics. The behavioral revolution in political science and sociology was accompanied by efforts to use "power" and "influence" as central organizing concepts in the study of political relationships. Lasswell and Kaplan's *Power and Society,* published in 1950, was a major effort to build a framework for political inquiry with special reference to power. They defined power as "participation in the making of decisions" (p. 75). A decision was defined as a policy, or a course of action with reference to others, involving "severe sanctions (deprivations)" (p. 71). A power relationship was specified as one where "G has power over H with respect to values K if G participates in the making of decisions affecting the K-policy of H" (p. 75). The exercise of influence was defined more broadly as "affecting policies of others" (p. 71). The emphasis in the study of power relationships was upon *power over.*

No one would deny the potential for some to exercise power over others within the context of governmental institutions. The question is whether that possibility is a virtue or a vice. Is it possible to discipline the exercise of power so that the use of extreme sanctions by some to impose deprivations upon others can be minimized? This is why people have struggled to discipline the exercise of power through constitutional arrangements for distributing authority among diverse decision structures. *Differentiated structures* are the means for ordering human actions so that actors are required to take account of the interests of others. Varying structural conditions might, thus, be expected to affect varying degrees of "power over" or "power with" relationships (Follett, 1951).

Many behavioral scientists would respond that questions of virtue and vice are normative questions outside the scope of a positive science. Power and dominance, they would contend, are the facts of human existence. Science must be concerned with facts not values.

If, however, choice exists about *constituting* or *structuring* power relationships, we need to examine the considerations which inform such choices. Calculating the probable consequences that flow from alternative arrangements requires scientific knowledge informed by an understanding of causes and effects. Alternatives can be selected only in light of such calculations. Choice always has references to arraying alternative possibilities and selecting from the array.

CRITERIA FOR CHOICE

Studying power relationships, thus, requires that we ask the questions: What purpose does power serve? Why do homo sapiens have reference to power relationships? In the revolutionary rhetoric of the American Revolution, it was contended "that government was instituted to promote the welfare of mankind, and ought to be administered to the attainment of that end" (Declaration of the Causes and Necessity of Taking Arms, July 6, 1775). According to this statement, the welfare of mankind is the criterion to be used for choosing among alternatives structures of power relationships. But, unless those choices are made using a warrantable body of knowledge about the effects of alternative structures of power relationships, such aspirations would still leave people in Rousseau's paradox where men are born to be free but are everywhere in bondage.

The term welfare is a summary term composed of many elements. The elements comprising the summation refer to the array of goods and services that contribute to improvements in the welfare potential of any individual or community of individuals. In the most general sense the criterion of efficiency implies improvements in *net* welfare. Improvements in net welfare can accrue only where all benefits, both monetary and nonmonetary, exceed all costs in relation to any form of human action. The rule of *Pareto optimality* attempts to account for both individual and collective welfare by specifying that net improvements in welfare occur only when changes in welfare either leave everyone better off or no one worse off (Buchanan, 1962; Buchanan and Tullock, 1962, chaps. 7, 13).

The criterion of Pareto optimality is clouded with controversy over the relative justness or injustice reflected in existing distributions of wealth as claims to goods and services. Criteria of efficiency, thus, need be accompanied by considerations of justice, fairness, or equity (Rawls, 1971). Multiple criteria of choice might be used to select from among available alternatives where both allocational and distributional consequences are calculated on the basis of estimates about probable cause-and-effect relationships.

Welfare considerations can thus be disaggregated to determine the effects of alternative decisions regarding the supply of various goods and services and the institutional structures that are appropriate to arrange for their production, delivery, and consumption. By shifting the focus from power relationships to service delivery systems we focus not upon dominance as the essential characteristic to be studied but upon the effects that are produced by different forms of collective action. This is not inconsistent with Lasswell and Kaplan's conception of power as participation in the making of decisions.

DESIGN, IMPLEMENTATION, AND EVALUATION

Power relationships are not ends in themselves. They are tools or instruments that can be assessed by their contribution to human welfare. As such they are subject to design, implementation, and evaluation.

The design of decision-making structures always has reference to stipulations in law which assign decision-making capabilities among persons occupying positions in relation to some specifiable community of interest that will yield some product, service, or contribution to human welfare. Positions are always specified in relation to some structure of relationships which serves some purpose (Commons, 1959, chap. 4). Any design will be based upon expectations about the relationship of structural conditions to consequences. Such expectations are a priori formulations and thus have standing as theory (Gregg, 1975). Whether or not such expectations or theories are warrantable—yielding the anticipated consequences—is problematic in the sense that all theory is problematic. The test for the warrantability of theory is whether the explanation holds in practice.

Whether structures make a difference can be established by examining varying decision-making arrangements for their effect upon performance with reference to service delivery. Ideally, similar types of services might be examined to hold production technologies as constant as possible. The delivery of water services, for example, involves quite a different production technology than the delivery of educational services. A focus on service delivery implies that the interface between service agencies and the population served permits comparisons across a variety of different organizational arrangements. Efficiency considerations apply to the choice of organizational structures as well as to production technologies. Our concern is whether structures affect performance in the delivery of services when measured in light of explicit evaluative criteria.

THEORY OF PUBLIC GOODS

People have long been aware that the nature of goods affects calculations bearing upon human welfare. Aristotle, for example, observed: "For that which is common to the greatest number has the least care bestowed upon it" (Aristotle, 1943, book 2, chap. 3, p. 83). Until the last two decades little explicit attention had been focused upon the characteristics of public or collective goods as distinguished from private or individual goods. An extensive literature has developed on the theory of public goods (Samuelson, 1954; Olson, 1965; Mishan, 1969). Several attributes have been identified which have substantial implications for the organization of decision-making arrangements. In this discussion I focus upon the attributes of (1) exclusion, (2) jointness of consumption, and (3) measurement to illustrate some of the considerations that have important implications for organizational arrangements.

EXCLUSION

Exclusion has long been identified as a necessary characteristic of goods and services that can be supplied under market conditions. Exclusion occurs when a potential user can be denied access to a good or service unless he meets the terms and conditions of a vendor. If both agree, a good or service can then be supplied at a price. An exchange occurs; the buyer acquires the good; and the seller acquires the value specified.

Where exclusion is infeasible, any one individual can derive benefits from a good so long as it is supplied either by nature or by the efforts of others. The air we breathe can be viewed as a good supplied by nature where exclusion is infeasible. Air, noise, and water pollution are instances of welfare-impairing events where individuals cannot exclude themselves from bearing the burdens of consuming noxious elements. Reducing the noxious quality of such pollutants is the equivalent of securing a public or collective good.

JOINTNESS OF USE OR CONSUMPTION

Another attribute of a good or service pertains to its jointness of use or consumption. No jointness of consumption will exist when consumption by one person forecloses its use or consumption by another person. Jointness of consumption implies that the use or enjoyment of a good by one person does not foreclose its enjoyment by others. Samuelson had specific reference to this attribute when, in his article on "The Pure Theory of Public Expenditure," he defined "a public good as a good which is subject to joint or collective consumption where any one individual's consumption is non-subtractible from any other individual's consumption of a good" (Samuelson, 1954:387).

Few, if any, joint consumption goods are perfectly nonsubtractible. The use and enjoyment of gravity as a force which firmly keeps our feet on the ground may be an illustration of a case of perfect nonsubtractibility. Most joint consumption goods are subject to partial subtractibility where at certain thresholds of supply one person's use of a good may subtract *in part* from its use and enjoyment by others. Congestion then begins to occur. Each increase in use impairs the use of the good for each other person in the community of users. Highways, for example, become subject to congestion where adding more users causes delay for all users. Public goods can thus be subject to degradation or erosion in the quality of the service unless supply is carefully proportioned to demand (Buchanan, 1970). A "tragedy of the commons" is likely to occur (Hardin, 1968).

Jointness of consumption is an attribute that is independent of exclusion. Some goods are subject to joint consumption but may be subject to exclusion when access can be denied to the relevant domain. A theater, for example, may allow for joint enjoyment or "consumption" of a performance, but access to the theater is subject to payment of an admission fee. A walled city can also attain exclusion by controlling admission to those who wish to reside, enter, or do business within the city. Those within share many amenities and annoyances jointly because individuals within cannot be excluded. Even in the unwalled city, boundary conditions establishing territorial jurisdiction may be a way of distinguishing between residents and nonresidents in the provision of some services—education, for example—which are available for use by those within the boundaries. A form of partial exclusion may exist in such circumstances.

MEASUREMENT

Since public goods are difficult to package or unitize, they are also difficult to measure. Measures usually cannot be translated into cardinal numbers so that gross production can be calculated like tons of wheat or steel. Qualitative measures such as the amount of dissolved oxygen in water, temperature of air, or decibels of noise can be used to describe important characteristics of goods subject to joint consumption. But such measures cannot be aggregated in the same way that gross production can be calculated for a steel factory. The task of measuring performance in the production of public goods will not yield to simple calculations but depends instead upon estimates where social indicators or various proxy measures are used as crude measures of performance. Private goods are easier to measure, account for, and relate to cost-accounting procedures and management controls.

SOME IMPLICATIONS

In his *Logic of Collective Action* (1965), Mancur Olson demonstrated that public or collective goods, defined as goods subject to joint consumption where exclusion is difficult or impossible to attain, present serious problems in human organization. If a public good is supplied by nature or the efforts of other individuals, each individual within the relevant domain is free to take advantage of the good since he cannot be excluded from its use or enjoyment. A cost-minimizing individual would have an incentive to take advantage of whatever is freely available to him without paying a price or contributing his proportionate share of the effort to supply a public good. So long as rules of voluntary choice apply, some individuals will have an incentive to "hold out" or act as "free-riders," taking advantage of whatever is freely available. If some are successful in pursuing a holdout strategy, others will have an incentive to follow suit. The likely short-run consequence is that voluntary efforts will *fail* to supply a satisfactory level of public goods.

Market institutions will thus fail to supply satisfactory levels of public goods and services. Because exclusion is infeasible, recourse to some form of collective action where sanctions can be mobilized to foreclose the holdout problem and compel each individual to assume his proportionate share of the burden will be necessary. In small groups, individuals may be relatively successful in keeping account of each other's efforts and coercing each other to assume one's proportionate share of the effort to procure jointly used goods. Families as a consequence can cope with potential holdouts and provide joint consumption goods under reasonably satisfactory conditions. But large groups are less successful in coping with the provision of public goods shared by a whole community of people. Each individual is more anonymous. Each person's share of the total good may seem insignificantly small, and each can function as a holdout with greater impunity. Potential recourse to coercion in levying taxes and foreclosing holdouts will be more important. This is the reasoning behind Aristotle's contention that the good or property shared in common by "the greatest number has the least care bestowed upon it."

Patterns of organization that can mobilize coercive sanctions are necessary for the operation of a public economy. This is why people seek recourse to governmental institutions. The provision of law and order is simply one of many public goods that are important to the welfare of human societies. Market institutions will fail to supply such goods and services because markets are characterized by exclusion, exchange, and voluntary transactions. Thus, the theory of public goods allows us to anticipate that some structures will fail and other structural conditions will be necessary to the operation of public service delivery systems. Structures, in this sense, do make a difference, and no one type of structures is equally good for all purposes (Ashby, 1962).

But recourse to coercive sanctions and governmental organization is not a sufficient condition for the delivery of public goods and services under relatively optimal conditions. Instruments of coercion can be used to impose deprivations upon others and dominate the allocation of values in a society. Furthermore, difficulties in measuring the output of public goods and services imply that governmental officials will also have problems in monitoring the performance of public employees. Management of public enterprises will be subject to even less effective control than the management of private enterprises, where outputs can be measured in quantifiable units.

THE ORGANIZATION OF
PUBLIC SERVICE DELIVERY SYSTEMS

The supply of reasonably satisfactory levels of public goods and services requires recourse to some form of governmental authority where sanctions can be mobilized to foreclose potential holdouts. How do we conceptualize structural arrangements that are appropriate to public service delivery systems? The traditional approach is to have recourse to a theory of sovereignty, use governments as the basic unit of analysis, and conceptualize an integrated administrative hierarchy of public functionaries as the appropriate structure for organizing public service delivery systems. In the course of the last decade this traditional approach has been challenged. The rudiments of an alternative approach are being developed by a number of political economists. We shall review each of these approaches. Each posits relationships between structure and performance, and each provides a different set of conceptual tools for interpreting evidence about the performance of urban service systems.

GOVERNMENT AND BUREAUCRACY: THE TRADITIONAL APPROACH

Much of the 20th century political analysis begins with a definition of the state as a monopoly over the legitimate use of physical force in a society. Such a definition of the state carries with it the presumption that all states are essentially unitary in nature. Some ultimate center of authority is presumed to exist that has the "last say" in settling disputes and in making decisions for society as a whole (Dahl, 1976).

The maintenance of a monopoly over the legitimate use of force in a society presumes that the necessary means for realizing such a monopoly must be available to those who exercise the prerogatives of government. The concept of unity of command implies that all of the means necessary for the maintenance of peace and security of a state must be available to those who exercise ultimate authority. Max Weber conceptualized a *monocratic structure* as one where "all functionaries are integrated in a hierarchy culminating in a single head" (Rheinstein, 1967:334). Unity of command is thus associated with an elaboration and perfection of bureaucratic administration, and bureaucratic administration is conceived as an ideal model for organizing the delivery of public services in a modern state.

Bureaucratic Administration

Weber contends that bureaucratic organization is technically superior to all other forms of organization for creating a system of public administration (Gerth and Mills, 1946:214). Bureaucratic rule according to Weber emphasizes an "objective" organization of conduct according to calculable rules without regard to persons. Bureaucracy has a rational character; rules, means, ends, and matter of factness dominate its bearing. Thus, bureaucracy involves a technical application of calculable rules of law to factual situations in a logically rigorous and machine-like manner. A bureaucracy through the uniform application of rules provides the instrumentality for controlling public service delivery systems so that uniform standards of service are supplied throughout a political domain.

The Paradox of Bureaucratic Administration

A radical shift in Weber's assessment of bureaucratic organization occurred when he characterized some of the social and political consequences which he associated with "perfection" in bureaucratic organization. "Where the bureaucratization of administration has been completely carried through," Weber anticipated that "a form of power relationship is established which is virtually indestructible." It is an instrument that can be "easily made to work for anybody who knows how to gain control over it" (Gerth and Mills, 1946:228).

When viewed from the perspective of the individual bureaucrat, the virtual indestructibility of the perfected bureaucratic machine implies, in Weber's words, that:

> The individual bureaucrat cannot squirm out of the apparatus in which it is harnessed. . . . He is only a single cog in an ever moving mechanism which prescribes for him an essentially fixed route of march. [Gerth and Mills, 1946:228]

Weber noted that altering the course of conduct in a bureaucratic machine normally depends upon the initiative of those at the very top. However, he went on to indicate the powerlessness of those at the top:

Under normal conditions, the power position of a fully developed bureaucracy is always overpowering. The "political master" finds himself in the position of the "dilettante" who stands opposite the "expert," facing the trained official who stands within the management of administration. [Gerth and Mills, 1946:232]

Weber contended that this powerlessness of the "master" holds whether the master is a "people," a "parliament," an "aristocracy," a "popularly elected president," or a "monarch" (Gerth and Mills, 1946:232-233).

Weber's characterization of bureaucracy as an ideal type stands in contrast to his assessment of the consequences which follow when "the bureaucratization of administration has been completely carried through" (Gerth and Mills, 1946:228). Can a "scientific" sociology or political science be built upon such a radical gap between ideal-type models and the consequences which are associated with their "perfection" or "full development"? So far as I know, Weber never attempted to resolve the anomoly implied by his conclusions about the "full development" of his "ideal" type.

The Anomoly or Paradox as a Problem of Institutional Failure

In the last two decades some important advances have occurred in the study of bureaucratic organizations. A number of sociological studies have focused upon a variety of bureaucratic dysfunctions including tendencies toward goal displacement and risk avoidance. The rule of action, "when in doubt, don't," stands in marked contrast to Weber's presumptions about speed, dispatch, and technical efficiency. Michel Crozier's study of *The Bureaucratic Phenomenon* has provided us with a study of French bureaucracy that reveals patterns of behavior more consistent with Weber's characterization of the fully-developed bureaucracy rather than his ideal type (Crozier, 1964).

In his *Politics of Bureaucracy* (1965), Gordon Tullock reformulated a theory of bureaucracy which eliminates the anomoly inherent in Weber's work. Instead of assuming that bureaucratic functionaries are engaged in a rational application of legal reasoning to factual situations, Tullock assumes that bureaucracies are composed of self-interested individuals who advance their careers by seeking promotions to higher positions. Tullock further assumes that career advancement depends upon the favorable recommendations of superiors. As a result of these two assumptions, Tullock anticipates that career-oriented functionaries will select strategies where they will repress information that leads to an unfavorable assessment by superiors and will facilitate a flow of information that leads to favorable assessments. This filtering process will systematically distort the information available to superiors. The longer the chain of command the more distorted information becomes. Distortion of information will diminish control and create expectations that diverge from events generated by actions. Large-scale bureaucracies will thus become error-prone and cumbersome in adapting to rapidly changing conditions. Rhetoric about organizational purposes and goals will diverge from performance. Goal displacement will occur. Efforts to correct the malfunctioning of bureaucracies by tightening controls will magnify errors. A decline in return to scale can be expected. After exhausting scale economies, the larger an organization becomes, the smaller the percentage of its activities will relate to output and the larger the proportion of its efforts will be expended on management and control systems.

Given the failure of information and control channels in large-scale bureaucracies, we would expect those in central authority to develop multiple command structures rather than relying upon a single hierarchy of authority. The French system of prefectoral administration represents a redundant command structure apart from the other technical and administrative services concerned with education, public works, public utilities, police, and other specialized services. The Soviet administrative system relies upon redundant hierarchies of authority represented by the party hierarchy, the state hierarchy, and the secret police apparatus.

Those assuming entrepreneurial responsibility for the operational performance of a given agency will thus be confronted with the task of acting in relation to multiple centers of authority, and we can expect subordinate units to engage in an increasing degree of "bureaucratic free enterprise" (Tullock, 1965:167). Side payments become increasingly important in the supply and delivery of public services. The rule of law is eroded by the availability of dispensations. Those who cannot afford dispensations bear the burden of encumbrances and immobility (Tocqueville, 1955). Corruption becomes the controlling fact of public life.

Implications for Local Government Organization

The full force of the anomaly or paradox inherent in the theory of bureaucratic administration has been avoided in many countries where local governments are maintained as self-governing institutions responsible for the delivery of many of the public goods and services necessary for everyday life. Instead of a single integrated hierarchy of public functionaries reaching from the center of government to the neighborhood, intermediate structures of political authority maintain distinct public service systems that are accountable to residents of the local community through democratic institutions of local control. Where strong traditions of local autonomy have existed historically, as in Switzerland or Great Britain, local institutions have been the basic structures responsible for the delivery of urban public services.

However, in recent decades, an increasing number of public services have been nationalized in response to the changing patterns of interdependencies inherent in both national and international economic and social relationships. Since the Great Depression and World War II, instrumentalities of national administration have become increasingly important deliverers of service even in those countries, like Great Britain, that have had a strong tradition of local self-government.

Increasing nationalization of public service delivery systems has been accompanied by extensive efforts to reorganize local units of government. The dominant approach in these reorganization efforts has been to view municipalities as miniature states subject to the same principles of organization as a nation-state but in a subordinate position to sovereign authorities operating at the national level.

The quest for a single center of authority also informs the search for an appropriate form of local government organization. Thus, the following principles have guided reform efforts to improve the performance of local government by increased efficiency and economy.

- Each major urban area should be organized into only one unit of local government.

- Voters in each major urban area should elect only the most important policy-making officials, and these should be few in numbers. Citizens will be confused by long ballots and be frustrated in their effort to choose among candidates for numerous offices or for numerous jurisdictions. Few officials are more visible and can be subject to more effective popular control.

- Local administration should be organized into an administrative command structure in accordance with hierarchical principles where authority tapers upward and culminates in a single chief executive who is accountable to the elected policy-making officials functioning as the citizens' representatives.

- The administrative functions should be separated from electoral politics. The work of administration should be performed by professionally trained public servants who are adequately compensated and employed on a full-time basis (Anderson, 1925:641-642).

The application of these principles results in proposals to eliminate small units of local government and consolidate them into a single integrated monopoly that can serve the larger regional community of interest associated with the growth of urban regions. From this perspective overlapping jurisdictions imply multiple delivery systems with duplication of services. Duplication of services implies waste and inefficiency in government. Efficiency is presumably enhanced by eliminating the smaller jurisdictions and by consolidating all authority into one jurisdiction responsible for the government of each major urban area.

The consolidation of local units of government into large urban monopolies may be subject to the same dynamics of institutional weakness and failure that Tullock anticipates for large bureaucratic organizations (Savas, 1971). The problem is compounded when these megamonopolies produce a large array of different public goods and services. Such services are not subject to easy measurement. In the absence of output measures the upper echelons of management are not capable of monitoring performance. The size of organization may exceed the bounds of efficient teamwork and generate grossly inefficient performance as employees pursue career opportunities while serving their "masters" at the upper echelons of the bureaucracy rather than their fellow citizens.

Instead of proceeding on an assumption that all governments exercise a monopoly over the legitimate use of force in a society, we might inquire whether the long traditions of municipal liberties and local self-government provide an alternative basis for conceptualizing public service systems as something different from fully integrated monopolies. If we return to the theory of public goods we can begin to conceptualize alternative structures for public service delivery.

PUBLIC GOODS AND COLLECTIVE ORGANIZATION

Instead of beginning with a theory of sovereignty, using governments as the basic unit of analysis and deriving conclusions about the structure of authority relationships, public-choice economists begin with individuals as the basic unit of analysis. They assume that goods are scarce and that different individuals may attach different values to goods and services. They stipulate a set of decision rules that structure opportunities and constraints for individuals to act in relation to one another. They then reason through the implications that follow when representative individuals choose strategies that enhance their net well-being. The two approaches—traditional political analysis and public-choice analysis—share common theoretical foundations reflected in much earlier works based upon the use of methodological individualism by Thomas Hobbes, David Hume, Adam Smith, Alexander Hamilton, James Madison, and Alexis de Tocqueville, among others.

The theory of public goods plays a key role in public choice analysis—particularly in the assessment of alternative arrangements for the organization of public service systems. In providing public goods, governments, like households, can be viewed first as collective-consumption units. Once the collective-consumption aspects of collective organization have been identified we can then turn to the production side. Government can also be viewed as producers concerned with the supply and delivery of public goods and services (V. Ostrom et al., 1961). We shall distinguish between these two functions by referring to *collective-consumption units* and *production units.* A single unit of government may, in fact, include both types of organization within its internal structure. The theory of public goods, within this context, leads one not only to focus upon different problem sets but to derive different implications flowing from alternative production-consumption arrangements (Bish, 1971; Bish and Ostrom, 1973).

Collective-Consumption Units

The problems to be solved in the organization of collective-consumption units are to foreclose the holdout problem and provide arrangements for levying taxes or coercing user charges from all beneficiaries comprising the collective-consumption unit. Relying upon the threat of coercion to levy taxes implies that little or no information will be derived about demand for a public good or service. Payment of taxes indicates only that a taxpayer prefers paying his taxes to suffering from penalties for nonpayment. As a consequence, alternative mechanisms to prices need to be developed for articulating and aggregating demands for public goods.

An appropriately constituted collective-consumption unit would have boundary conditions and a territorial domain that includes within its jurisdiction the relevant beneficiaries who share a common interest in the good or service subject to joint consumption. That collective-consumption unit will be constituted to take operational decisions under decision rules of less than unanimity. This is necessary to foreclose a holdout position. It will hold a limited monopoly position and require authority to exercise coercive sanctions, but it need not exercise a monopoly over the legitimate use of force for a society as a whole. Voting rules, modes of representation, and rules applicable to the taking of operational decisions about taxes, expenditures, and levels of public goods or services to be procured will need to be formulated as means to supply essential information about preferences and how those preferences can be translated into a decision about collective demand. If decisions can be made under a set of decision rules where benefits exceed costs and costs can be fairly proportioned among beneficiaries, each individual would have an incentive to agree to such a form of collective organization, forgo holdout strategies and procure a joint consumption good (Buchanan and Tullock, 1962).

Production Units

A production unit, by contrast, would be one which can aggregate the technical factors of production and yield a flow of goods and services to meet the requirements of a collective-consumption unit. The organization of an appropriate production unit would include authorization of a manager who would assume entrepreneurial responsibility for aggregating factors of production and organizing and monitoring performance of a production team that would supply the appropriate level of the good or service for a sum that covered costs. Problems arise, however, in proportioning supply to patterns of use.

Proportioning Supply for Joint Users

The characteristic of *partial subtractibility* of consumption implies that increased use or alternative uses may impair the value of a good or service for other users or other uses. Use of air to discharge wastes, for example, may impair its use for other purposes. Some uses may drive out other uses and lead to a serious erosion in the quality of urban life (Buchanan, 1970). Jointness of use under conditions of partial subtractibility may require rules for ordering patterns of use so as to reduce potential conflict among the community of users. If rules are to be effective in ordering patterns of use, mechanisms for their enforcement must be available. Unless those rules take account of varying patterns of use and supply conditions in discrete circumstances, they are likely to become serious impediments to joint well-being.

Coproduction

Another problem in proportioning supply to patterns of use arises when users of services also function as essential coproducers. The quality of educational product, for example, is critically affected by the productive efforts of students as users of educational services. The health of a community depends as much on the informed efforts of individual citizens to maintain good health as it does upon the professional services of health delivery systems. The peace and security of a community is produced by the efforts of citizens as well as of professional policemen. Collaboration between those who supply a service and those who use a service is essential if most public services are to produce the intended effects (Whitaker, 1976).

Professionalization of public services can be accompanied by a serious erosion in the quality of those services. This is especially true when professionals presume to know what is good for people rather than provide people with opportunities to express their own preferences, and when they fail to regard citizens as essential coproducers of many public goods and services. Higher expenditures for public services supplied exclusively by highly trained cadres of professional personnel may contribute to a "service paradox," where the better that services are, as defined by professional criteria, the less satisfied that citizens are with those services. An efficient system of public administration will depend upon professionals working under conditions where they have incentives to assist citizens as essential coproducers rather than assume that citizens are incompetent to realize their own interests.

Choice of Production Units

Problems associated with partial subtractibility of goods and services subject to joint use or consumption and coproduction may lead a joint-consumption unit to organize the supply of a public good or service through its own production unit. In that case the joint-consumption unit and the production unit would serve the identical population. Yet, the constitution of the two units may be essentially separable. The common council representing the joint-consumption unit may confront complex bargaining with managers of production units to procure an appropriate supply and delivery of public goods and services for a given level of expenditure (Niskanen, 1971).

Apart from the problems created by partial subtractibility and potential conflict among joint users and coproduction, the essential problems of nonmarket provision and production of public goods and services can be resolved through the organization of appropriate collective-consumption units. A collective-consumption unit may face a variety of production possibilities for securing the supply and delivery of an appropriate quantity and/or quality of a public good or service.

The members of a common council for a collective-consumption unit might decide to contract with a private vendor to supply a public good or service. In that case they are required to translate decisions about the desired quantity or quality of public goods or services into specifications that can be used to secure bids from potential vendors, state the terms and conditions for contractual arrangements, and establish standards for assessing performance. Such a common council also needs to employ a purchasing agent to negotiate with potential vendors, receive service complaints from service users, and monitor a vendor's performance in delivering the stipulated quantity and/or quality of service. A variety of municipal services including street sweeping, solid waste collection, fire protection, engineering, planning, construction of public works, educational services, among others, can be supplied by private vendors.

Another possibility is for a collective-consumption unit to contract with a production unit associated with a different unit of government. A municipality acting as a collective-consumption unit may contract with another municipality or some other unit of government to supply police services, fire services, educational services, or a wide range of other public services.

A fourth possibility exists when a collective-consumption unit decides to rely upon its own production unit to supply some components of a service but to rely upon other production units to supply other components of a service. Its own production unit may draw upon other producers to supply it with factors of production, serve as a purchasing agent to procure the delivery of supplemental services, or as a joint producer supplying a mix of services rendered by the joint effort of multiple production teams. Any given collective-consumption unit may rely upon the joint production efforts of several different producers in supplying and delivering a particular bundle of goods and services subject to joint consumption.

As soon as we begin to array some of these alternatives, a wide variety of possibilities for organizing public service delivery systems become apparent. A system may include a large number of autonomous units of government with substantial degrees of overlap among multiple levels of government. Each citizen participates in multiple consumption units organized to realize diverse communities of interest through overlapping levels of government and is served by an array of different producing units. These arrangements can be viewed as *public service industries* composed of many collectivities rather than by a single monopoly producer.

Much of the regulatory capability in such a public service economy can be attained through patterns of interorganizational arrangements. If such interorganizational arrangements are subject to error-correcting and efficiency-inducing tendencies, we have reason to believe that a compound public-service economy would perform better than a highly monopolized or unitary system. If, conversely, the patterns of interorganizational arrangements are dominated by high external costs, threats, and counterthreats, we would expect a compound system to be subject to substantial error and to serious erosion of individual and collective welfare.

Interorganizational Relationships in a Compound System

Proportioning Consumption and Production Possibilities. In a world where goods subject to joint consumption vary in domain from household to global proportions, the availability of an array of differently sized collective-consumption and production units will provide opportunities to realize diverse economies of scale. Where heterogeneous preferences for public services exist, advantage may be gained by having relatively small collective-consumption units. As long as a collective-consumption unit can articulate preferences for its own constituency and has access to a reasonably equitable distribution of income, the collective-consumption unit can specify the mix of services preferred, procure an appropriate supply, and provide compensation. In this case a small collective-consumption unit might contract with a large production unit and each might take advantage of diverse scale considerations in both the consumption and production of a public good or service.

Another circumstance may exist where the appropriate collective-consumption unit is large but efficient production is realized on a smaller scale. The appropriate consumption unit for users of interstate highways in the United States, for example, is probably national in scope. This national unit functions as a "provisioner" by developing appropriate specifications and financial arrangements for procuring interstate highway services. However, variability in climatic and geographic conditions over a large continental area are probably such that production and maintenance can be more efficiently supplied by smaller organizations. Thus, the U.S. Department of Transportation acts as a buyer of interstate highway services from state highway departments and private firms which act as the principal production units. Similarly, elementary education is typically produced in relatively small units—elementary schools—but the social benefits may accrue to a much larger community.

Where charges can be appropriately levied on individual beneficiaries, use taxes can substantially alleviate the problems associated with rationing the use of a joint good when partial subtractibility results in potential congestion costs (Mushkin, 1972). Highway construction and maintenance, highway police patrols, and other services for motorists could, for example, be charged against gasoline taxes rather than other forms of general taxation. User charges or use taxes lead beneficiaries to calculate the cost of a service as against the value of its marginal use. Reliance upon rules and criminal sanctions to order the use of a public good or service that is made freely available to all users may place undue emphasis upon criminal sanctions if user charges can more appropriately proportion use to the level of demand.

Competition, Bargaining, and Cooperative Efforts. If each collective-consumption unit has potential access to several production units and is prepared to consider alternative options in arranging for the supply of a public good or service, the relationships between collective-consumption units and production units will take on the characteristics of a quasi-market relationship. The market in this case is *not* between producers and *individual* consumers. We would expect such market structures to fail. The quasi-market, instead, arises in relationships among collective-consumption and production units.

If the potential producers include an array of private vendors and public agencies, an opportunity exists for bargaining to procure public goods or services at least cost. Such bargaining can be expected to yield increased efficiency in a public-service industry. The opportunity for bargaining among collective-consumption units and production units also creates incentives on the part of the bargaining parties to increase levels of information and to develop indicators of performance. Decisions made with access to better information about potential benefits, costs, and performance would be expected to generate error-correcting propensities in inter-organizational relationships.

Bargaining may also occur in a noncompetitive situation where multiple production units may be able to gain a joint benefit by coordinating their actions with one another. Joint efforts may be extended to organizing supplemental enterprises to supply a variety of indirect services such as crime laboratories, police training academies, and joint dispatching services. Where high levels of interdependency have been developed through cooperative arrangements, collective-consumption and production units can be expected to develop regularized organizational arrangements to reduce bargaining costs. These arrangements often take the form of a voluntary association with regularly scheduled meetings and with officials to set meeting agendas and to arrange for the organization and presentation of pertinent information.

The essential decision rule in such cooperative arrangements is one of unanimity. Any one collective-consumption unit or production unit can refuse to go along and is not bound by the decision of the rest. Willingness to compromise usually depends upon the costs of invoking some alternative governmental decision-making arrangement such as courts or legislatures. If the prospects of an imposed decision would be worse than what members of an association can work out for themselves, they will have an incentive to compromise interests and seek a unanimous solution. In such situations, self-regulating tendencies are derived from mutual agreements.

Conflict and Conflict Resolution. Cooperative arrangements maintained under a rule of unanimity can always be threatened by the presence of a holdout where multiple collective-consumption or production units are creating significant externalities for one another. If those externalities have the characteristics of a public good in a larger domain that impinges upon several collectivities, some one collectivity may find it advantageous to hold out and enjoy whatever benefits it can derive from the joint actions of others without assuming its proportionate share of the costs. If some holdouts are successful in their strategy, others will follow suit, cooperative arrangements will fail, and the "tragedy of the commons" will work itself out with an erosion in welfare for everyone concerned. The maintenance of a holdout strategy and the impending threat of tragedy may lead some to respond to holdouts with threats or counterthreats. Unless constrained by the availability of institutions for adjudicating and resolving conflicts, threats and counterthreats can escalate into violence and warfare (Boulding, 1963).

A highly compounded political system *without* substantial overlap among the many jurisdictions is especially vulnerable to this form of institutional failure. Americans in their ethnocentrism refer to this dynamic as "Balkanization." Michel Crozier also referred to this phenomenon as the "vicious circle of decentralization"–a characteristic form of institutional failure which he identified with the American political system (Crozier, 1964:236).

With overlapping units of government, conflicts among governments at any one level may be resolved by recourse to the decision-making arrangements existing at a higher level of government. Such arrangements are inherent in federal systems of government. The critical consideration is the availability of legal, political, and constitutional remedies to the parties injured as a consequence of negative externalities generated by governmental action at any given level of intergovernmental relationships. If such remedies do not exist, then the task of fashioning political solutions through constitutional decision-making processes will need to be considered. In a sense, Europe has been undertaking such a task with the organization of the European community. In time, the availability of an appropriate set of institutions in the European community may facilitate the efforts of many urban communities to work out joint arrangements to procure the delivery of urban services on a multinational basis.

COPING WITH ALTERNATIVES, IMPLICATIONS, AND CHOICES

The existence of alternative conceptions and theoretical explanations for the operation of public service delivery systems provides the essential point of departure for a new generation of scholarship and for the design of different organizational structures (V. Ostrom, 1974). The first task in developing traditions of political inquiry and experimentation is to learn to use language as a tool for reasoning through the implications that follow from specifiable assumptions and postulates. The task of clarifying the logical implications that flow from stipulated conditions must be accompanied by efforts to test the logic to determine whether the implications hold when applied in practice. The structures of reasoning used to derive inferences can be stated as hypotheses, and explanations can be submitted to evidentiary tests. The logical sufficiency of an analysis needs to be supplemented by empirical tests to establish empirical warrantability. The availability of multiple theoretical explanations can be the source of multiple hypotheses. The theory that offers the better explanation in accounting for relationships among variables is the more useful analytical tool. Unless political rhetoric can be disciplined by theoretical criticism and evidentiary tests, we cannot develop a language for intelligent political discourse.

Efforts to apply this mode of analysis to studies of electoral systems, coalition formation, the organization of interest groups and political parties, patterns of fiscal relations, bureaucracies, and other forms of collective action will all contribute to an understanding of basic relationships inherent in public economies. However, such elements always function as components in a complex ordering of political-economic relationships. Scholarship that focuses upon specialized decision structures thus will need to be complemented by contextual analyses concerned with complex orderings of interdependent decision structures. All public economies involve simultaneous games where the pursuit of strategy in any one game has implication for the play of other games.

Empirical analysis of public service systems will need to function at many different levels of inquiry using different elements and units of analysis. For some purposes individuals operating within a particular type of structure may be the appropriate level of analysis. For other purposes it may be necessary to view units of government, public agencies, and private firms as elements operating within a particular public service industry supplying a particular type of public service to specifiable communities of people (Gregg, 1974). Yet the logic underlying such empirical analysis needs to be based upon a presumption that units of government, public agencies, and private firms are all composed of *individuals* who relate themselves to one another through specifiable rule structures ordering activities related to specifiable payoffs. The structure within units of government, public agencies, and private firms will vary from one another. If they all function within a particular public service industry, we might anticipate that the game within each separate structure will also be affected by the interorganizational game that is played among the different elements within an industry (E. Ostrom et al., 1974). Furthermore, we might anticipate that the industry structure for the police industry will be vastly different in Switzerland than it is in Italy or France, for example.

If social scientists develop a warranted body of knowledge that indicates how structures affect performance, their work will have substantial relevance to those who are called upon to function as decision makers in one capacity or another. Whatever conception is acted upon can be regarded as a hypothesis, and the actions taken can be viewed as an experiment to test the conceptions being acted upon (Dewey, 1927; Campbell, 1969). Organizations are simply efforts to design ways of evoking information, ordering choices, and structuring actions. The forms of organization that best facilitate learning are the ones most likely to yield the best performance in the long run.

In *Democracy in America,* Tocqueville contended that peoples in democratic societies would require a new science of politics to guide them in governing their own affairs (Tocqueville, 1945, vol. 1, p. 7). He referred to this new science of politics as a "science of association" (vol. 2, p. 110) which would enable men to "comprehend the utility of forms" (vol. 2, p. 325) and put the doctrine of self-interest to proper use as a rule of action for organizing collective enterprises (vol. 1, p. 10). Such a science of association would specify the relationship of structures to performance and enable men to advance the art of associating together. By returning to a concern for structure and performance, perhaps we can begin to contribute to this "new science of politics."

REFERENCES

ANDERSON, W. (1925). American city government. New York: Henry Holt.

ARISTOTLE (1943). Politics. New York: Modern Library.

ASHBY, W.R. (1962). "Principles of the self-organizing system." Pp. 255-278 in H. von Foerster and G.W. Zopf (eds.), Principles of self organization. New York: Macmillan.

BISH, R.L. (1971). The public economy of metropolitan areas. Chicago: Markham.

BISH, R.L., and OSTROM, V. (1973). Understanding urban government: Metropolitan reform reconsidered. Washington, D.C.: American Enterprise Institute for Public Policy Research.

BOULDING, K.E. (1963). "Towards a pure theory of threat systems." American Economic Review, 53(May):424-434.

BUCHANAN, J.M. (1962). "The relevance of pareto optimality." Journal of Conflict Resolution, 6(December):341-354.

——— (1970). "Public goods and public bads." Pp. 51-71 in J.P. Crecine (ed.), Financing the metropolis. Beverly Hills, Calif.: Sage.

BUCHANAN, J.M., and TULLOCK, G. (1962). The calculus of consent: Logical foundations of constitutional democracy. Ann Arbor: University of Michigan Press.

CAMPBELL, D.T. (1969). "Reforms as experiments." American Psychologist, 24(April): 409-429.

COMMONS, J.R. (1959). Legal foundations of capitalism. Madison: University of Wisconsin Press.

CROZIER, M. (1964). The bureaucratic phenomenon. Chicago: University of Chicago Press.

DAHL, R.A. (1976). Modern political analysis (3rd ed.). Englewood Cliffs, N.J.: Prentice-Hall.

DEWEY, J. (1927). The public and its problems. New York: Henry Holt.

DYE, T.R. (1976). Policy analysis: What governments do, why they do it, and what difference it makes. University: University of Alabama Press.

FOLLETT, M.P. (1951). Creative experience. New York: Peter Smith.

GERTH, H.H., and MILLS, C.W. (eds., 1946). From Max Weber: Essays in sociology (Galaxy Book ed.). New York: Oxford University Press.

GREGG, P.M. (1974). "Units and levels of analysis: A problem of policy analysis in federal systems." Publius, 4(fall):59-86.

——— (1975). Problems of theory in policy analysis. Lexington, Mass.: D.C. Heath.

HARDIN, G. (1968). "The tragedy of the commons." Science, 162(December):1243-1248.

LASSWELL, H.D., and KAPLAN, A. (1950). Power and society: A framework for political inquiry. New Haven, Conn.: Yale University Press.

MISHAN, E.J. (1969). "The relationship between joint products, collective goods, and external effects." Journal of Political Economy, 77(May/June):329-348.

MUSHKIN, S. (ed., 1972). Public prices for public products. Washington, D.C.: Urban Institute.

NISKANEN, W.A., Jr. (1971). Bureaucracy and representative government. Chicago: Aldine-Atherton.

OLSON, M. (1965). The logic of collective action. Cambridge, Mass.: Harvard University Press.

OSTROM, E., PARKS, R., and WHITAKER, G.P. (1974). "Defining and measuring structural variations in interorganizational arrangements." Publius, 4(fall):87-108.

OSTROM, V. (1974). The intellectual crisis in American public administration (rev. ed.). University: University of Alabama Press.

OSTROM, V., TIEBOUT, C.M., and WARREN, R. (1961). "The organization of governments in metropolitan areas: A theoretical inquiry." American Political Science Review, 55(December):831-842.

RAWLS, J. (1971). A theory of justice. Cambridge, Mass.: Harvard University Press.

RHEINSTEIN, M. (ed., 1967). Max Weber on law in economy and society (Clarion Book ed.). New York: Simon and Schuster.

SAMUELSON, P.A. (1954). "The pure theory of public expenditure." Review of Economics and Statistics, 36(November):387-389.

SAVAS, E.S. (1971). "Municipal monopoly." Harper's Magazine, 243(December):55-60.

TOCQUEVILLE, A. de (1945). Democracy in America (2 vols.; P. Bradley, ed.). New York: Alfred A. Knopf.

——— (1955). The Old Regime and the French Revolution (Doubleday Anchor Books ed.). Garden City, N.Y.: Doubleday.

TULLOCK, G. (1965). The politics of bureaucracy. Washington, D.C.: Public Affairs Press.

WHITAKER, G.P. (1976). "Size and effectiveness in the delivery of human services." Paper presented at the annual conference of the American Society for Public Administration, Washington, D.C.

Part II

COMPLEX ORDERINGS

2

The Compound Structure of
Public Service Delivery Systems
In Israel

DANIEL J. ELAZAR

□ MOST RECENT LITERATURE on urban public service delivery systems has been generated in the United States and has addressed American conditions. A considerably smaller body of literature has been emerging in Europe, particularly the Scandinavian countries, and is addressed to conditions in the various states on that continent. By and large, theory and analysis in the field reflect these locational factors. Hence, any widening of the circle of such studies can only add to the possibility of improving our ability to do comparative analysis and thereby our understanding of the problems and possibilities of urban public service delivery systems.

In this respect, Israel offers many advantages for advancing knowledge in this field. Like the United States, Israel is a new society, founded "from scratch" as a result of the migration of self-selected populations to "virgin" territories (that is, territories perceived by the migrants to be essentially uninhabited at the time of their settlement). In both countries, settlers underwent a frontier experience as a part of the process of settlement and had to covenant or contract consciously with one another to create social and political institutions (Elazar, 1973b; Hartz, 1964; Lipset, 1963; Turner, 1962). At the same time, Israel is a new society whose founders did not come from the British Isles or Northern Europe as did those of most of the other new societies extant in the world today. The great majority of Israel's settlers came from Asia or Eastern Europe, the Near East and North Africa, bringing with them very different cultural baggage (Eisenstadt, 1967). It may well represent the only new society founded by people of those cultural backgrounds.

The Jewish community of prestate Israel was referred to as a "new society" by its members and was built through a series of interlocking covenants and compacts in the manner of all new societies. The construction of the modern Israeli polity began in the last generation of the 19th century. The first modern Jewish agricultural settlement was established in 1878. By World War I, when that generation came to an end, the well-known second *aliya* (literally, "ascent"—the term used to describe prestate waves of migration to the country) of 1903-1914 had brought in the nucleus of the subsequent founders of the state who created the first national institutions that gave the state its tone (Lacqueur, 1972). The founders were for the most part socialists imbued with 19th century socialist ideologies as these were expressed in the socialist circles of Central and Eastern Europe. Because of this, covenants and contracts were oriented toward a cooperative, rather than an individualistic, model of social and political organization with a strong collectivist tinge. This orientation was reinforced by the Jewish political culture of the founders. As a result, when the state of Israel was established as an outgrowth of the Jewish Yishuv (settled community) of the prestate period, it assumed very extensive responsibilities within the polity. Today Israel's polity consists of a society extensively permeated by governmental activity. Thus Israel may be the only new society developed on the basis of socialist principles.

Israel has an emergent political culture that contains a number of conflicting elements yet to be sorted out and integrated (Elazar, 1971a). Principal among these are (1) a statist-bureaucratic political culture which implicitly accepts the concept of a reified state existing independently of its citizens and which views political organization as essentially centralized, hierarchical, and bureaucratic in character (a view shared by the vast majority of those Israelis coming from continental Europe), (2) an Oriental subject political culture, which views government as the private preserve of an elite, functioning to serve the interests of that elite and hence a potentially malevolent force in the lives of ordinary people (a view shared by the great majority of the Israeli population coming from the subject cultures of Eastern Europe and the Arab countries of the Middle East and North Africa), and (3) a Jewish political culture, which is civic and republican in its orientation, viewing the polity as a partnership of its members who are fundamentally equal as citizens and who are entitled to an equal share of the benefits resulting from the pooling of common resources. This culture combines a high level of citizen participation with a clear responsibility on the part of governing authorities to set the polity's overall direction and is shared to a greater or lesser degree by the 85% of the population that is Jewish.

These three political cultures exist in somewhat uneasy tension with one another. This tension is evident in a great gap between the formal institutional structure of the polity (which is an expression of European statism) and the actual political behavior and informal institutional arrangements which make it work. Formally, Israel is a highly centralized, hierarchically structured bureaucratic state on the model of France. In fact, the state and its institutions function on the basis of myriad contractual agreements which assume widespread power sharing on a noncentralized basis. These are enforced through a process of mutual consultation and negotiation in which every individual party to an agreement must be conciliated before action is taken.

Because Israel is still an emergent society, the precise political-cultural synthesis cannot yet be forecast. So, for example, in 1975 the proportional representation, party-list electoral system, which has been a feature of modern Israel since the beginning of the Zionist effort, was modified to provide for the direct election of mayors independently of their city councils and to endow them with a modest veto power over council actions. This radical departure represents a step away from continental European parliamentar-ianism toward a separation of powers model which is more consonant with Jewish political culture.

Finally, Israel is an exceptional phenomenon in the world of modern territorial states in that it is intimately linked to the Jewish people, an entity with political characteristics not confined to a particular territory. Israel itself has indeterminate boundaries, a condition which is presented to the world as a product of momentary circumstances but which has been characteristic of the Middle East since the dawn of recorded history. Moreover, a great part of its political life is not territorially based but is rooted in confessional, consociational, and ideological divisions at least as permanent. It is not that Israel is aterritorial, but territory is only one of the dimensions which its people and institutions use in organizing space and time for political purposes.

In sum, the Israeli system offers special and unique opportunities for extending the comparative study of new societies, political culture, political development, and what may be new trends in the political organization of space in the postmodern era. What follows should be considered a prolegomenon to a more thorough study of public service delivery systems in Israel, the first fruits of a larger research project being undertaken by the Institute of Local Government at Bar-Ilan University under the author's direction. This project involves studies of local influence on state activities within local communities, the diffusion of innovations among local governments, the impact of political subcultural differences on local service delivery systems, and the delivery of specific services.[1]

THE COMPOUND STRUCTURE OF THE ISRAELI POLITY

For those familiar with Western European, and most particularly American institutions (where polities are territorially based, where government is organized fairly simply on two or three levels or planes, and where the greatest complexity is found in the overlapping of local governments), the Israeli situation is complex indeed. For those familiar with the American federalist theory of the compound republic (Ostrom, 1971), it is of particular interest to note that the Hebrew word used to describe the organization of a polity or government *(leharkiv)* means "to compound." The same word is used to describe complexity, offering etymological testimony to the expectations inherent in the environmental and cultural matrix in which the Jewish people always have been imbedded and in which Israel functions today. The fundamentally contractual character of Jewish political life is reflected in the idea that bodies and polities are compounded from different entities that retain their respective integrities in the larger whole (Elazar, 1971b).

ISRAEL AS A JEWISH STATE

The state of Israel can be seen as a republic compounded in a variety of ways. First of all, it is, by self-definition, a Jewish state. This means not only a state with a Jewish majority, but one in which the Jewish people as a corporate entity can express its particular culture, personality, and values and which seeks to foster the expression of that triad as, perhaps, its principal task. It is in this respect that many Israelis, including the leadership, consider the state a part of a larger entity known as the Jewish people. Israel is the only politically "sovereign" state within that entity and, as such, occupies a unique position. It is also, for certain purposes, a Jewish community and maintains relationships with other Jewish communities on what could be considered a federal basis. The Jews of Israel, particularly the most politically conscious among them, see the fostering of this relationship as one of the tasks of the state—indeed, a task that does not escape local government either.

The principal institutional manifestations of this special relationship between Israel and the Jewish people are found in the "national institutions" functioning within the state's territory. These institutions are so named because they are considered to belong to the entire Jewish people (in Zionist terminology, nation) and not to the state of Israel alone. Among these are the Jewish Agency and the World Zionist Organization (WZO), which are responsible for land settlement in Israel and the Zionist education of Jews in Israel and outside, and the Jewish National Fund (JNF) which is responsible for land purchase and reclamation throughout the country.[2]

The Hebrew University is also a national institution. Its library is the national library and is so named. The university's board of governors is drawn from the Jewish community worldwide; some two-thirds of its budget comes from world Jewish sources and only 10% from the funds of the state of Israel. Budgeting and policy-making powers are shared by the state's Council for Higher Education, the university's "national" governing board, and the university senate, composed of all full professors on full-time appointment. These are roughly the equivalent of state, federal, and local bodies, if one were to translate them into modern political terminology. While the Hebrew University is the only one formally designated by law as a national institution, all other universities in the country have the same status, de facto, since they have the same arrangements for governance and funding.

The Israeli government also seeks to institutionalize the relationship between Israel and the Diaspora Jewish communities through common organizations and associations, encourages study programs in Israel, and sends emissaries to Jewish communities overseas to work with them in strengthening Jewish life. Many of these study programs involve Diaspora Jews playing personal roles in agricultural settlements where they benefit from the services which those settlements have to offer.[3] Similar study programs in the cities are also subsidized by the Jewish Agency. Many cities subsidize institutes for higher religious study *(yeshivot)*, which draw Jews from the Diaspora as well. The local governments do this in line with the political culture of the state wherein one of the tasks of local government is viewed as fostering culture, personality, and values of the national population within each locality, Jewish or Arab.

Through the Jewish Agency and its related organizations, the Jewish people as a whole undertake numerous settlement, social, and educational projects throughout the land of Israel, in both rural and urban areas and often in cooperation with the local authorities. The various bodies have regional offices in different parts of the country and, in some cases, local ones as well, which serve local populations in their spheres of authority as if they were governmental agencies. In addition, the Jewish Agency is principally responsible for financing the construction of such local facilities as high schools and community centers with funds raised outside Israel.

Finally, the Law of Return, which guarantees every Jew (except those fleeing criminal prosecution of one kind or another) the right of entry into Israel and more rapid naturalization than non-Jewish immigrants, in effect obligates the state and local governments of Israel to provide services to every Jewish immigrant from the moment of his or her settlement in the country.[4] In fact, because of the dominant political culture, such services and benefits are extended immediately to all those accepted as residents of the state, without regard to ethnic, national, or religious distinctions.

ETHNORELIGIOUS PLURALISM IN ISRAEL

On another level, the state of Israel is compounded of several different ethnoreligious minorities in addition to the Jewish majority: Muslim Arabs (344,000); Christians, mostly Arab (77,300), divided into various churches; Druse (38,000); Bahai; Circassians; and Samaritans (480)—each with its own socioreligious structure and legal status. Following the Middle Eastern pattern, all of these groups seek to preserve their corporate identity, and Israel has granted them a legal status and institutional framework through which to do so. While the legal status and institutions of each are adapted to its particular situation, with Muslim Arabs having the most comprehensive and the Circassians the least, all have certain basic institutions and government support for their activities as corporate entities as well as the normal services provided to all citizens.

Because of the pattern of settlement common in the rural Middle East, whereby villages are either ethnically homogeneous or shared by perhaps two ethnic groups, local government becomes a major vehicle for the expression of these corporate interests. The Israeli government has made great efforts to encourage villages housing these minorities to acquire full municipal status and to utilize the instrumentalities of local government not only to provide local services but to express the cultural personalities and values of the groups within them. In a sense, this represents a partial adaptation to the realities of what, in the period of the Ottoman Empire, was known as the millet system, whereby every ethnic group was constituted as a millet with its own internal autonomy.[5]

In sum, Israel is a republic compounded of different ethnoreligious groups, each recognized and supported by the state, yet claiming its own higher source of authority. Since, in Israel, personal status matters are by law the province of the religious communities (this too, is a common Middle Eastern pattern), every person must be a member of some religious community if he or she hopes to get married, divorced, or buried. Of course, while an individual may choose to use only these minimal services from his or her religious community, the communities provide many more services and are expected to do so by the faithful attached to them.

As a result, the various religious communities have institutional structures of their own, which are recongized in law and which in some cases are governed by bodies chosen under state law because they provide state-supported services and thus must follow certain standard procedures with regard to selection and representation. Each religious community has its own religious courts, whose judges are supported by the state, hold commissions from the state on the basis of qualifications determined by each religious community, and are selected by the appropriate bodies of each religious community under procedures provided for by state law (Jaffee, 1966). These courts administer the religious laws of their communities—each of which has its own legal system for matters within its competence. Religious laws stand in relationship to the secular legal system in Israel roughly as state laws stand in relationship to federal law in a federation with a dual legal system.

The principal administrative organs of the religious communities vary from community to community. The Christian communities have no separate administrative bodies, other than the church hierarchies, which also handle matters of religious law, because they are essentially in the Catholic tradition and also play a more limited role in the lives of their citizens. The religious functions of the Muslim communities are administered through the Wakfs, the Muslim religious trusts; and in the Jewish communities, every locality with a Jewish majority has a local religious council consisting of laymen and rabbis elected through a complex formula and partially supported by local government. All of these bodies are, in some respects, responsible to the Ministry of Religions, whose minister is a member of the cabinet and which is the channel through which state funds reach the various religious groups.

From the point of view of the state, these religious groups obtain their powers through state law. From the perspective of each of the religious communities, however, their powers flow directly from Heaven and their law represents the Divine will. As far as they are concerned, the state has only a minimal role in determining their existence and certainly no legitimate role in determining their powers other than that to which they are willing to acquiesce.[6]

Each of the several communities represents a further compound within its ranks. Every Arab locality is a compound of extended families—really clans—so much so that voting and political office-holding, not to speak of decision making and the distribution of political rewards, are dependent upon competition or cooperation among the extended families in each locality. Every so often a group of young people emerges to challenge this arrangement, and there is talk that the Arabs are modernizing and will no longer be bound by this kind of familial loyalty. However, all but the most radical of the young usually end up following the lead of their families in these matters.

THE COMPOUND OF IDEOLOGICAL PARTIES

The Jewish community in Israel is a compound based upon federal connections between different Zionist movements. Various groups of socialist Zionists, each with their own ideology, erected their own settlements and institutions in the country. Paralleling them were Zionists with a liberal (in the European sense) ideology and others whose primary ideology was derived from traditional religion. The latter groups ranged from religious socialists who based a modern collectivist ideology on ancient religious sources to the religious right who saw no reason to allow any kind of secular thinking or behavior in the state to be.

Each of these movements sought to create as comprehensive a range of institutions as it could, a kind of nonterritorial state of its own, but within the framework of the overall Zionist effort. Since they also wanted the overall effort to succeed, they federated together through various roof organizations and institutions through which they could pursue the common objective of a Jewish homeland, even while contesting with one another with regard to the shape of the state to come and the vision that would inform it. This federation of movements became the basis fo the present party system which organizes and informs Israel's political system.

Today, the country divides into three "camps": labor, liberal or center, and religious (with the last closer to labor than to the center in most respects). (See Eisenstadt, 1967; Fein, 1967.) The remarkable stability of voting patterns in Israel since 1948 (and since 1935 if the elections to governing bodies of the prestate Jewish community are included) is a reflection of this basic division. Such electoral shifts as have taken place rarely have crossed the boundaries of the camps, reflecting only changes within each. Even the masses of post-1948 immigrants who tripled the population of the state were settled, employed, educated, and politically absorbed on the basis of the "party key" through which the relative strength of the various parties within the three camps was maintained.

At one time, virtually all services for Jewish citizens were provided through the parties or, in the case of labor, through the Histadrut, the General Labor Federation, which united several different labor parties for certain purposes. Again, the analogy to a federal system is apt.[7] Just as in a federal territorial polity one has to be a resident of a state to avail oneself of the services of the polity as a whole, so too in prestate Israel was it necessary to be a member of a party. With the establishment of the state, the government took over more and more of the services, beginning with the military services (before 1948, the movements actually had separate paramilitary formations), continuing with the schools and most social services. The parties still retain control of sports, health insurance and ordinary medical facilities, and, to some extent, banking. Even those functions that have been absorbed by the formal institutions of government maintain an informal division by party for employment purposes.

The importance of the compound of parties is such that even the most casual student of Israeli affairs is aware of it. Manifestations of the old divisions are, however, disappearing. More and more services are provided neutrally by the state or local governments or, as is more often the case, through cooperative arrangements involving both. Party influence exists in the government structure and primarily touches those who pursue governmental careers rather than the public at large. Only in the religious camp do the ideological justifications remain sufficiently strong to create demands of prestate intensity. These are accommodated by allowing for parallel institutions in many fields.

THE COMPOUND OF ECONOMIC SECTORS

A final means by which Israeli society is compounded is through what is known as the "sectors," whereby public activity is divided between distinctive governmental, cooperative, and private sectors along one plane and urban and rural sectors along another. As a government-permeated society, it is not surprising that most enterprises in Israel exist by virtue of government assistance, either through direct investment, loan guarantees, or simply sponsorship with appropriate tax benefits, favorable foreign exchange rates, and the like. While the Israeli government has veered heavily away from the strongly socialistic position of its leadership at the beginning of statehood, this has not meant any movement toward laissez-faire. Government's role remains as great as ever in promoting state-permeated social capitalism within a mixed economy. It is enhanced by the fact that Israel's precarious security situation and narrow economic base give political decisions preference over economic ones in most cases.

If the government sector is invariably the strongest, the cooperative or workers' sector is the oldest and most hallowed. It emerged in the 1920s, when the various small collectives of the Jewish pioneers were brough together to create the Histadrut and the Hevrat Ovdim (Wokers' Society), which was to be the means through which the Labor Federation could establish and maintain its own complex of economic activities. While the cooperative sector has diminished greatly in importance since the establishment of the state, it remains the biggest nongovernmental owner and operator in the country. The largest industries, including the largest conglomerate in the country, are under its ownership. The public transportation companies, with the exception of the state railroads and the miniscule internal airline which provide very limited service, are workers' cooperatives. The kibbutzim and moshavim are integral parts of the cooperative sector. The largest department store chain is Histadrut-owned. The Histadrut operates Kupat Holim (Workers' Sick Fund), the largest prepaid medical service in the country, which serves over a third of the population and, through it, a network of hospitals, and old-age and rest homes—in other words, the complete apparatus of a socialized health system. The local Kupat Holim clinic is one of the vital service centers in the local community. All this is in addition to being the great comprehensive union of most workers in the country.

The cooperative sector consists primarily of producer cooperatives as exemplified by the bus companies, construction companies, kibbutzim and moshavim. Even the nominal consumer cooperatives do not operate on the basis of Rochdale principles since their profits are funneled back into the Histadrut and not to their patrons. Moreover, the Histadrut itself has become considerably bureaucratized over the years, so that, while it has sought to serve the Zionist mission of rebuilding the state and the public good as its leaders have interpreted that good, it has also become at least as distant from its own members as the government is from the man in the street in most of its operations. This needs to be emphasized so as to gain proper perspective on the workings of that sector.

Finally, there is the private (meaning capitalist and small business) sector, which is growing after being the weakest of the sectors for many years. Although there has always been private enterprise within the Zionist effort, it was not a very acceptable form of pioneering activity, because the overwhelming majority of the pioneers were socialists. Only in recent years has the importance of a healthy private sector, both as a countervailing power to the other two sectors and in its own right, come to be recognized and even encouraged by the government. It is typified by the range of private entrepreneurial activities characteristic of all modern Western societies.

By and large, these three sectors have not developed on a competitive basis; rather, they tend to cooperate with each other as could be expected in a small country with a relatively weak economy and a political culture which emphasizes partnership as does that of the Jewish majority. Many enterprises are jointly developed by two or all three sectors. Efforts on the part of private investors—usually from overseas—to "buck" these arrangements have generally come to naught since, without a favorable government attitude, it is very difficult to succeed in any economic enterprise in Israel. What is particularly important is that the government, which is generally assumed to be in the business of providing services, also has a strong economic stake in the society, while the general labor association, which in other countries would be an interest group, plays an even more important role as a supplier of services on the "retail" as well as "wholesale" level and as an economic developer.

The division between urban and rural sectors is equally important. The rural sector tends to be self-contained in many respects, because almost all agricultural production and rural life is embraced within the framework of the kibbutzim and moshavim. Moreover, the special place which kibbutzim and moshavim occupy in the Zionist enterprise makes them the arbiters of the rural sector. (The original Zionist vision placed heavy emphasis on Jews returning to the soil where they would redeem themselves through cooperative toil in a natural setting.)

Not only are the kibbutzim highly integrated political, social, and economic units, the exact antithesis of fragmented urban society, but in many respects they are more modern than Israeli cities in their culture, behavior, and technological development. Every kibbutz is organized as a cooperative society in which all except the most personal items of property are legally held in common. Social life is necessarily intimate, with a common dining hall and other facilities to enhance the already great likelihood of high social integration that exists in any community of a few hundred to a few thousand population.

The kibbutz has municipal status as a Vaad Mekomi (local committee) under state law. It is actually governed by two principle bodies, the general meeting (equivalent to the American town meeting), which elects the local committee on a yearly basis and which meets monthly to consider major issues, and the local executive committee, which meets as frequently as necessary, sometimes daily, to deal with current business. Most of the day-to-day business of the kibbutz is carried on through a multitude of committees involving as many members as are capable of participating. Every kibbutz is also a member of a Moetza Azorit (regional council), a federation of contiguous settlements that provides secondary local government services, in which it is represented by a delegate or delegates chosen by its own general meeting.

The moshav is slightly less integrated than the kibbutz (Baldwin, 1972). In the moshav every family has its own family farm and private life with some work and all major purchasing and marketing done in common. This makes for a cooperative rather than a collectivized atmosphere, but since, like the kibbutz, the moshav is also small (moshavim are usually much smaller than kibbutzim), it tends to be a highly integrated social unit. Under the law, the moshav is both a cooperative society with shared economic functions and a municipal unit with its own general meeting and local committee. Moshavim are also members of the regional councils along with the kibbutzim.

Because of the particular character of rural settlement in Israel whereby even family farms are concentrated in villages with their own local institutions, the 728 rural settlements with their own local governmental autonomy have an average population of under 800. Moreover, rather than being very limited-purpose local governments, such as those serving populations that small would be in the United States, the kibbutzim and the moshavim provide comprehensive economic and social services as well as traditional municipal functions on a level that far exceeds almost anything to be found outside the Communist bloc.

The kibbutzim and moshavim offer many opportunities to test the effectiveness of public services delivery by small local governments operating under a variety of social and cultural conditions. While very little research has been done from this perspective, two points are clear. First, in a self-selected population (which is what these settlements represent) it is possible for these small communities to provide a very high level of services. Even so, it has apparently been increasingly necessary to increase the scale through which certain services are provided—hence, the growing power of the regional councils. All but the smallest settlements, for example, choose to maintain their own elementary schools, but the provision of an adequate high school requires a somewhat larger population base. Hence, the provision of high schools is increasingly entrusted to regional councils (Criden and Gelb, 1974). At the same time, it should be noted that the regional councils themselves are relatively small, ranging in population from 678 to 20,378, with only four over 10,000.

Because these rural settlements can bring to bear a full range of options—political, economic, social, and commercial—to confront any problem, they are the most autonomous local governments in the country and also the ones with the most effective cooperative arrangements with one another and with the state authorities. The greater internal diversity of the cities and their more limited corporate purposes prevents them from functioning nearly as well. Moreover, since cities are considered to be mere by-products of the Zionist movement, which, as a back-to-the-land movement, was in many respects anti-urban (Cohen, 1970), they do not have the same claim on the resources or respect of the state that the rural settlements do.

The cities are open to greater permeation by the external society—including the institutions of the state and the cooperative sector—in every respect. While the kibbutzim and moshavim are actually part of the cooperative sector, as the elite elements of that sector they can manage their relationships with it. Cities, on the other hand, are often dependent upon decisions taken by the cooperative sector at the higher echelons of its bureaucracy, over which they have minimal influence.

LOCAL GOVERNMENT: FORMS, NUMBER, AND SIZE

FORMS

Urban government in Israel legally takes two forms, with the distinction between them minimal (Meljon, 1966). The largest local communities are legally cities with full municipal powers, but, in the English tradition of *ultra vires,* they possess only those powers specifically granted to them, and, in the case of conflict with the state, city powers are interpreted narrowly. Small urban places are formally termed local councils, a status which gives them almost as much power as cities and in a few cases more, but which makes them more dependent on the Ministry of Interior for hiring personnel (Adler, 1960). Both kinds of municipalities are governed by councils elected on the basis of proportional representation in which the voter casts his vote for a party list rather than for individual candidates, and each party gets the number of seats reflecting the percentage of the total vote it garnered. Frequently, no party gains a majority and a coalition is formed to govern the city, much as is the case on the state level. In some cases, even parties winning a majority will form coalitions in order to strengthen the hands of the local government or to better distribute local political rewards in consideration of statewide coalitions (Weiss, 1969; Elazar, 1973a).

While cities and local councils are the basic urban municipal units, they can federate with one another to create larger, special-purpose municipal bodies designed to undertake specific tasks. These bodies, termed federations of cities, can be established by two or more municipalities and can undertake one or more functions. They range from the Lod-Ramle joint high school district to the federation of cities of the Dan region, which encompasses the better part of the Tel Aviv metropolitan area and provides several functions which seem to be best handled on a metropolitan-wide basis.

Israel also has utilized the equivalent of special districts for certain purposes. In Israel, these are called authorities. By and large, these authorities handle water drainage and sanitation problems which require adaptation to watersheds that are less conveniently adapted to existing municipal boundaries. The local religious councils in the Jewish-dominated localities, local planning committees, and the state-mandated, quasi-independent local agricultural committees established in most former agricultural colonies that have become urbanized are kinds of special districts also.

The cooperative sector is represented locally by local workers' councils which are elected by vote of all members of the Histadrut within each council's jurisdiction (which, in most cases, more or less conforms to the municipal boundaries). While formally private, many of their activities are of a quasi-governmental character, and they usually wield great political influence. These workers' councils play a role somewhat equivalent to that played by a chamber of commerce in a small American city. The fact that workers' councils play a role in Israel similar to that played by businessmen's associations in the United States is a significant indicator of Israel's unique political history and culture.

NUMBER

There are today a total of 1,297 local authorities functioning in Israel, or approximately one local government per 2,315 inhabitants as compared to a ratio of one to 2,650 in the United States.

Table 1 summarizes the kinds of local authorities functioning in Israel and the number of each. By any standard, this is a high figure. It is particularly high given the strong formal commitment in Israel to centralized government, both in terms of state-local relations and within localities.

Americans take fragmented local government as a matter of course, since the American system has been designed that way from the beginning. Indeed, American reformers rail against the tradition

TABLE 1
LOCAL AUTHORITIES IN ISRAEL

Type	Number
Cities	35
Local councils	114
Regional councils	48
Local committees	728
Federations of cities	32
Religious councils	204
Agricultural committees	26
Planning committees	84
Drainage authorities	22
TOTAL	1,293

of local fragmentation as one which needs to be overcome and which would not exist in any rationally organized polity. In this respect, the Israeli experience is especially interesting. There, every conscious effort has been made to avoid the multiplication of local units and, in particular, the creation of overlapping units serving the same population. Yet the exigencies of objective reality as well as internal political considerations have created a condition of growing fragmentation which is approaching that of the United States. Indeed, it can be assumed that the only reason that it has not reached American proportions is that, in Israel, the state undertakes many functions directly which have been assigned to special districts in the United States.

SIZE

Another consideration in dealing with public service delivery systems in Israel is that most local authorities serve relatively small populations. Tel Aviv, once the largest city in the country and still the central metropolis, has a population of approximately 340,000 and is already on the decline, having peaked at approximately 385,000 a decade ago. It is now undergoing the process of dedensification which has become common in central cities over much of the Western world, as the movement to better housing in newer parts of the metropolitan area plus urban renewal with the construction of new housing at lower densities has had its impact. Jerusalem now has approximately 350,000 people and Haifa approximately 220,000. There is a second cluster of five cities with populations in the vicinity of 100,000 population. The other 141 cities range in size from 80,000 to 200. The average city size is under 18,000. Table 2 classifies Israel's cities by size category. Nearly half the population lives in villages or small cities of under 40,000 population (the dividing point between small and medium-size cities in the United States), while approximately 25% live in cities of over 200,000.

Moreover, neighborhoods have real meaning in Jerusalem and Haifa. In part, this is associated with the very formation of the cities themselves, whose modern founding was the result not only of associations of pioneers established by compact for that specific purpose, but also of a compounding of different neighborhoods, each created independently by a pioneer association and then linked through a second set of compacts to form the present city. Haifa, where formal neighborhood institutions are strongest and most widespread, reflects this process to the fullest. As each neighborhood merged with the growing city, it preserved a neighborhood committee with specific if limited responsibilities for the provision of services and for participation in the development of common city-wide services insofar as they affected it. Jerusalem was unified by external decision of the ruling power, but, because most of the older neighborhoods represented clearly distinct socioreligious communities, the city has consistently refrained from imposing itself upon them in those fields of particular concern to each. Today it, too, is trying to extend more formal devices for neighborhood participation to newer neighborhoods where other forms of distinction remain important.

In Tel Aviv the merger of neighborhoods was more thorough, and little, if anything, remains of the earlier framework other than names and recollections, but today the city is making some effort to revive consultative bodies in at least those neighborhoods which have preserved the most distinctive personalities.

TABLE 2

ISRAEL'S CITIES, BY POPULATION SIZE CATEGORY

Population Size	Number of Cities
200,000 +	3
80,000-149,000	5
40,000-79,000	8
20,000-39,000	12
8,000-19,000	33
4,000-7,900	32
2,000-3,900	29
Under 2,000	22

Israel's cities tend to be considerably smaller than most American reformers, including those advocating smaller rather than larger units, recommend. In Israel, as in other parts of the world, there is some pressure to consolidate small local units. Despite the fact that the Minister of the Interior has full authority to abolish any local unit or consolidate two or more units, this authority has rarely been used and then only when such a move has sufficient political backing from local elites. In the early days of the state when political elites did not include representatives of the localities in question, more consolidations were effected. In the last decade, however, even the weakest local governments have acquired political bases of their own, and any moves to consolidate would be strongly resisted. As a result, consolidation efforts have essentially ground to a halt to be replaced by efforts to create federations of cities to undertake those functions which the individual communities cannot undertake by themselves.

To date, the federation of cities device has been generally used to undertake functions of metropolitan concern and has been little used in the more rural parts of the country. This is partly because the federation of cities idea was developed to serve cities that adjoin one another, that is to say, those in metropolitan regions. The device has not been extended to free-standing cities within a region which may be separated by no more than a few miles but which see themselves, and are treated as, totally separate entities. Thus, a certain amount of very real intergovernmental collaboration in planning and service delivery has been developed in the Dan region (the accepted name for the Tel Aviv metropolitan area), which consists of some 20 cities whose boundaries are contiguous with one another. Yet in the Galilee, a region of several hundred thousand people with no single city of 40,000 population but with six cities of over 10,000 all within an area of less than 1,000 square miles, there are relatively few intermunicipal arrangements and little local concern with moving in that direction. This is true even though the region as a whole shares common state facilities (e.g., a large hospital in Safed, university extension courses in that city and near Kiryat Shmona, district offices in Nazareth, rudimentary sewage treatment facilities near Tiberias) and has the potential of becoming a kind of multinodal metropolitan region of the kind that has developed in such places as central Illinois and the Connecticut River valley.

PATTERN OF ORGANIZATION IN THE DAN REGION[8]

The metropolitan-technological frontier began to emerge as a factor in Israeli life by the 1960s. It began on the coast, primarily in the Dan region, extended northward and southward along the coastal plain, and is beginning to protrude into the interior. The Dan region remains its focal point. It has three principal nodes: Tel Aviv proper, which is the commercial and administrative headquarters for the enterprises of the new technology; the Lod district, which is the site of Israel Aircraft Industries, Israel's largest single enterprise and the principal manifestation of an industry based upon the new technology; and Rehovot, the site of the Weizmann Institute, which has attracted a number of small science-based industries. Each of these represents a classic pattern on the metropolitical-technological frontier.

As we have already indicated, Tel Aviv, the commercial center, is diminishing as a place of residence even as its position as a headquarters city is being strengthened. Those who work in the city's offices increasingly live in a chain of suburbs to the north and northeast of the city that have many of the characteristics of American bedroom suburbs. As more and more people acquire automobiles of their own, they commute from outlying areas, where they develop their own municipal institutions with most of the middle-class, reform-oriented politics characteristic of such suburban communities. This is a radical transformation in a country where political parties have dominated almost every aspect of life in the manner of other consociational democracies.

These municipalities are singularly antiparty in their orientation. Politics within them is dominated by local lists whereby candidates of the same socioeconomic background run for office with the promise of providing more extensive services more efficiently. These local lists have stimulated far more citizen participation than has been the norm in contemporary Israeli politics and have recruited nonprofessionals into political ranks. Once elected, these office-holders attempt to provide relatively inexpensive and efficient administration, the goal of which is service and in which normal political considerations play a minimal role.

The Lod node focuses on Ben-Gurion Airport, the country's major international airport, and not on any particular municipality. Israel Aircraft Industries is not associated with a particular city. Seemingly it is located out in the fields, though in fact it is within a local jurisdiction. Its workers come from all parts of the metropolitan region. It and other airport-related, science-based industries in the vicinity have stimulated the creation of two or three bedroom suburbs similar to those generated by Tel Aviv. In one, the city government is dominated by El Al flying personnel who have acquired houses next to the airport and have been attracted into local politics in order to develop and maintain a community to their taste. There, too, suburban-style politics is the norm.

The Rehovot node focuses on one of the original rural land settlements which has now reached a population of 46,400. Rehovot is perhaps the most successful town in its size range in the country. It is a classic example of a city that has capitalized on every aspect of the metropolitan-technological frontier. The Weizmann Institute gives it a major institution of higher learning, oriented toward science, both theoretical and applied. The institute not only has attracted a resident population of high caliber but has served to attract a large number of industries and other institutions that seek an academic-cum-scientific atmosphere. This in turn makes Rehovot a prestige address within the Dan region.

The city has also capitalized on its old agricultural connections to acquire a role in the new industrialized agriculture, both as a shipping point for agricultural produce and as the locus of the Faculty of Agriculture of the Hebrew University. Finally, Rehovot has captured the commerce of much of the southern part of the country almost as far south as Beersheba, despite the plans of state planners who sought to develop other regional nodes. These factors have led to the development of a substantial retail trading base plus many professional services. Despite extensive government efforts to encourage the development of other centers, Rehovot has captured the market in almost every retail and service field.

All this is reflected politically in the emergence of strong local candidates with a reform orientation. Since Rehovot is not a new city, the national parties are well entrenched on one level, and separate local parties have not emerged. On the other hand, the present mayor rose to power on a reform platform and has been able to eliminate national party influence to a substantial degree and transform the local party branch into an instrument reflecting his goals. Rehovot's city administration is known as a progressive one, on the forefront of governmental innovation in the country and highly service-oriented.

The overall structure of the Dan region reflects the patterns of the metropolitan-technological frontier. From the beginning, Tel Aviv was unable to become the dominant central city that emerged in the United States and Western Europe as a result of massive consolidation efforts that eliminated adjoining communities as independent political entities. Today Tel Aviv represents approximately one-fourth of the total population of the metropolitan region. Three other cities of over 100,000 share borders with it, and a fourth lies just beyond. The four together have a population some 90,000 greater than Tel Aviv proper. Another 400,000 people live in cities of between 30,000 and 100,000 population in the region, while some 200,000 more live in smaller places.

The region has three major institutions of higher education and a branch of a fourth, two of which are in Rehovot, one in north Tel Aviv, and the fourth in Ramat-Gan (population 120,000), making the region multinodal in that respect as well. Travel to work crisscrosses the region, with many people going from Tel Aviv outward to the exurban communities.

Every town in the region is quite distinct as a city in the minds of the inhabitants. Even people who live in a bedroom suburb will not identify themselves as being from Tel Aviv, but rather from their municipality. Since Israel has no county government or its equivalent and no other mediating government between the state and the municipalities (cities are not members of regional councils), there is an added structural incentive to local self-identification. The municipality is the provider of all local services either directly or through some federation of municipalities in which its own identity is clearly maintained.

FACTORS AFFECTING THE OPERATION OF
LOCAL PUBLIC SERVICE DELIVERY SYSTEMS

COOPERATION

Three major factors influence the operation of local public service delivery systems in Israel. The first of these is the effort at cooperative activity which has characterized Israeli society from the first. It reached its most intense form in the kibbutzim, whose experience seems to indicate that a shared response to the fundamental life-questions of religion and politics is utterly necessary in cooperative communities of this character, or the community will be faced with intolerable factionalism. In the early 1950s many kibbutzim actually split over questions of political ideology having to do with the extent of their socialist beliefs. These divisions led to secessionist movements and the literal division of certain kibbutzim into two independent cooperative societies with their own municipal institutions.

Each settlement is part of a chain of cooperative societies designed to provide some service or set of services which embraces the whole settlement movement in a series of ever-widening arenas of cooperative activity. The locus of control will vary from arena to arena depending upon the function involved (thus education remains a very local matter, while marketing is handled by the respective settlement movements). This network of cooperative institutional relationships is so strong that it weakens the degree to which regional councils can function as mechanisms for serving member settlements. The settlements often prefer to deal through their countrywide cooperative affiliates instead. In part, this tension is reduced by encouraging regional councils to form from the same settlement movement. Since the original pattern of settlement provided for clustering of settlements by movement, this is frequently possible, but in some cases it has led to gerrymandering. The direction in the settlement movements is toward more utilization of territorially based mechanisms to accomplish purposes which were formerly entrusted almost exclusively to each movement's Tel Aviv office.

Cooperative ties in the cities are far less intense. At their best, the cities become *civil* communities (communities organized for more limited civil or political purposes) and not comprehensive ones (Elazar, 1970). The few exceptions are small cities whose populations have a distinctive religious or ideological bent and, of course, the old established Arab municipalities, which are really traditional villages which have now acquired municipal status.

At the same time, even the cities can be understood as networks of cooperatives in at least one sense. Most people in Israel live in what Americans call condominiums and what in Israel are termed cooperative houses. The cooperative house represents an interesting merger of the exigencies of urban living with the cooperative orientation of Israeli society. Israelis have been encouraged to own their own homes. Today some 70% do. At the same time, there is a conscious government policy to encourage high density settlement in urban areas. Since land is controlled either by the government or by the Jewish National Fund, which works closely with the government, it has been able to make this policy stick. Thus most Israelis, except for those in the rural areas, the Arab villages, and a few bedroom suburbs, live in apartments.

The solution to the potential contradiction was to create a situation in which Israelis were essentially required to buy their apartments (there are almost no rental units available, and rental costs are so high that virtually no one rents more than temporarily). Thus every family has an undivided share in the commons of the building, and neighbors must cooperate with neighbors in the maintenance of those common areas and in the provision of common services. Where the building is legally a cooperative house, this is required by law, but, even where it is not, it is required by necessity. Thus, for the overwhelming majority of Israelis, the simple act of living requires cooperative links to control externalities. In the case of small buildings (up to eight families) it is likely that building governance will be in the hands of a committee of all resident adults, with one or more persons taking on specific responsibilities on a rotating basis, usually for one-year terms. In larger buildings, a committee is elected at an annual meeting of all tenants, whose responsibility then is to handle all but exceptional problems during its tenure, which is also usually a year. This arrangement follows the pattern of self-governing institutions in Israel and, indeed, is in the Jewish political tradition—that of the general meeting and the operating committee.

LOCAL ROOTS

A second factor affecting the operation of local public service delivery is that growing out of the historical pattern of the development of the state of Israel (Gutmann, 1958). Historical exigencies led to the state developing out of local roots. The rural-land frontier stimulated this on one level, with small groups of settlers coming together and organizing themselves locally to undertake pioneering tasks. The local role was further stimulated by the fact that the Turkish authorities who governed the land until 1917 saw their functions as essentially custodial and oriented to maintaining minimum security in the land; all else was left to the ethnoreligious communities to develop as they saw fit. The British authorities who came after the Turks (between 1917 and 1948) did not depart from this pattern except to make it more honest and efficient. It was left to the individual ethnoreligious communities within the country to determine the kind of public infrastructure that they wanted for themselves. For the ruling authorities, this was a natural and highly functional way to deal with the problem of differing ethnic groups with widely differing styles of internal organization and highly divergent expectations from the public sector.

With regard to the Arab villages, this policy meant that they remained almost entirely unchanged until the establishment of the state of Israel. The Arab leadership resisted municipalization for fear that it would interfere with their traditional dominance, so it came about only when a more educated generation emerged. The Israeli authorities encouraged them to acquire municipal status and the services and facilities that went with such status. As a result, the Arab villages have been undergoing modernization with regard to basic municipal functions for no more than a generation.

The Jewish sector, on the other hand, wished to push rapidly ahead with the development of a modern, Western-style society with all the public services that that entailed. Indeed, because of their socialist bent, the Jewish pioneers wished to provide even more services than many individualistic societies in the West. Since neither the Turkish nor the British mandatory authorities were interested in meeting their needs and since, for that matter, the Jews were not interested in having others do for them what they believed that they should do for themselves, the Zionist institutions undertook the task of providing those services. This task, to no small extent, fell upon the Jewish-sponsored local authorities which served most of the Jewish population.

Even local government law in the country was generally enacted by the British mandatory regime after the fact, i.e., after Jewish settlers had created local institutions which then had to be formalized. The regional councils are good examples of this. Jewish settlements clustered in various parts of the country found it useful and necessary to cooperate with one another for the provision of common regional services. Several such clusters—two on the coastal plain, two in the Jordan Valley, and two in the Jezreel Valley—began to do so in the 1930s, creating regional councils that had no legal status. In 1941 the mandatory government took note of those councils and promulgated a law providing for their recognition as formal local government bodies and for the establishment of others. Cities like Tel Aviv created statelike service systems for their residents under the permissive British rule and with the blessings of the Jewish National Committee. Public services of the new society thus had local roots and were pyramided into countrywide programs through various kinds of contractual and federal arrangements.

GOVERNMENT-PERMEATED SOCIETY

Finally, the fact that Israel is a government-permeated society strongly affects the local operation of public service delivery systems. Many services are provided by the Histadrut in the cooperative sector. These are closely linked to state services or policies. One of the major consequences of this is that local government officials must spend as much time working with outside authorities to either provide services or fund services as they do in directing their own affairs.

Increasingly, the situation of permanent crisis in which Israel finds itself has led to heavy strains on government budgeting with an inordinate share of the gross national product devoted to defense. As a result, local governments have been quite restricted in their ability to finance activities. Relatively few tax resources are at their disposal, and the local share of total governmental expenditures in Israel has been on the decline for nearly 20 years.

By and large, Israeli local governments manage to maintain their freedom of movement by managing deficits rather than through grantsmanship, with the former having become for them the functional equivalent of the latter. There are great restrictions on local government's taxing powers, but there are almost no restrictions on its borrowing powers, providing that any particular local authority can pay the high interest involved. Thus, local authorities borrow heavily from the banks in order to provide services and then turn to the state government to obtain the funds to cover the loans. As long as the services they wish to provide are in line with state policies (and there is almost universal consensus with regard to those services, so that this is not generally an issue) and there is some degree of unanimity within the local ruling coalition with regard to what is being done, the state will provide the requested funds. Nevertheless, this does mean that the local authorities must spend a very large share of their time in negotiations with their state counterparts.

Local leaders are also able to turn, in some matters, to the Jewish Agency or even directly to foreign donors to gain additional resources, mostly for capital investment—e.g., the construction of a new high school, a community center, or a child-care center. Where services are provided directly by the state, local authorities will use their influence to try to negotiate more and better services or to influence those responsible for delivering those services locally, but in this they are notably less successful than they are in mobilizing funds for their own programs, partly because the Israeli political culture encourages every officeholder to act as independently as possible.

DELIVERING SPECIFIC SERVICES

POLICE

The police force in Israel is an instrumentality of the state directly controlled by the Ministry of Police. All policemen are part of the central government police force, although every community of significant size has its own police station attached to it and in the course of time relations develop between the police officers stationed locally (most of whom are likely to live in the locality) and the local authorities. Nevertheless, it is fair to say that except in unusual circumstances the local authorities do not have any significant influence over the work of the police. The citizenry has even less, particularly since, while Israel generally has a tradition of maintaining the civil rights of individuals, there are few channels of citizen recourse for dealing with specific police violations of those rights. The greatest force enabling citizens to influence the police is a political culture which makes it possible for citizens to attempt to convince a police officer to change his or her line of action on grounds of justice or mercy, simply on the basis of bargaining and persuasion. This method tends to be most effective when "mercy" is involved, whether the matter involves a traffic ticket or the arrest of someone involved in a near-violent argument. The general tendency toward mercifulness in the local culture tends to emerge at such times and it is possible to play upon a police officer's sympathies.

The police force is generally considered efficient and effective and has experienced only a minimum of corruption over the years. In matters affecting security and the personal well-being of citizens, the police are generally considered responsive. In the matter of security, awareness of a common problem of the utmost gravity leads the police to spare no efforts to respond to citizen requests no matter how farfetched they might seem (one never knows whether a terrorist has left a bomb or whether the sack of garbage out of place is just that). Responsiveness to personal considerations arises from the network of interpersonal relations which is characteristic of Israeli society and which makes people wary of rejecting human overtures by failing to give a human response.

Police are considered less efficient, effective, and responsive in matters regarding burglary or the destruction of property. This is partly because property crimes are increasing and are much more difficult to deal with and partly because there seems to be an element in Jewish culture that views destruction of property as relatively unserious—particularly when compared to questions of life and health. The fact that this is a shared value allows the police more latitude in this regard, since a robbery victim is likely to feel that as long as he or she has come through unscathed (and Israeli thieves are not given to physical violence), matters are tolerable.

Summarizing the relationship between structure and performance, it can be said that since the police force is organized on a countrywide basis and not locally, smallness of scale does not seem to be an immediate factor in stimulting effectiveness, efficiency, or responsiveness. However, a deeper look would show that the overall characteristics of Israel, as a small society with strong interpersonal connections that cut across most social divisions, do function to prevent the impersonality that could arise and seems to have arisen in many large American local jurisdictions of similar size with regard to relations between police and public.

EDUCATION

Elementary and secondary education is provided by a partnership between state and local authorities in Israel. Israel does not have independent school boards. Instead, city councils handle whatever tasks are devolved upon them in school matters, generally through a vice-mayor for education and an education committee of the council. The state Ministry of Education funds all the operating costs of the regular elementary education program, the middle schools, and a few of the high schools. Teachers are certified and employed by the Ministry of Education.

Despite this apparently highly centralized structure, education in Israel in fact is rather decentralized. The local authorities are responsible for providing and maintaining school buildings and equipment (including texts, based upon ministry lists), managing the schools, and registering and enrolling the students and for virtually all ancillary and enrichment programs beginning with prekindergarten education. They also have direct control over almost all high schools in the country. Thus, the local departments of education are in a position to direct local educational affairs, and, since the ancillary and enrichment services are becoming an ever larger part of every school's program, their influence is expanding.

Matters are complicated by two other factors. The first is the division of public schools into several distinctive units. Within the Jewish community, there are separate state and state-religious schools, the latter embracing some 30% of the total student population within the state system, each with its own department within the Ministry of Education and within each local office of education. In addition, the state provides support (almost equal to that given state schools) to an independent school system for extreme religious elements who are not prepared to make what they perceive to be the compromises with Western secular civilization made by state-religious schools. Finally, the state maintains a network of Arabic schools for the Arabic-speaking minority.

Schools in the kibbutzim, while nominally linked to one of the two state systems, represent another subsystem because of the particular orientation of the kibbutz movements. Each of these subsystems has its own set of educational goals, which reflect strong religious, ideological, or cultural predispositions and which make them somewhat less than amenable to outside interference. In a political system in which pluralism has become consociational in character, their claims to autonomy are widely recognized. Furthermore, every school principal is virtually sovereign when it comes to matters within his sphere of competence.

Schools, like the rest of the state, were built from the bottom up, with parents and local branches of movements coming together to found individual schools before there was a central educational authority. The significance of this is compounded by the fact that every educational institution was designed to foster the values of the new society among the new generation, including whatever specific version of those values a particular school represented. As a result, virtually every school became a bastion of ideological as well as social and intellectual development, a key element in the creation of the new Jewish society. Principals and teachers were powerful figures —leaders in the struggle for national survival. Given the Jewish cultural predisposition toward treating learning and teaching with the utmost seriousness, this condition was even further intensified.

Once the state was established, it became inevitable that the schools would be welded together into a system, although the precise character of this system emerged only after a considerable political controversy in the early 1950s. While the schools formally had no choice in the matter, when they were compounded together to create the present system and subsystems, the principals and teachers were able to preserve many of their erstwhile prerogatives, formally, or informally.

Today the law provides that every principal has a right to change up to 25% of the curriculum established by the Ministry of Education for his school system. Given the extent to which certain subjects are commonly accepted as necessary, that percentage encompasses as much maneuverability as would be available even under an optimally flexible situation. To the best of this writer's knowledge, no regular school outside those in the kibbutzim has come close to exercising the full prerogative, although in recent years a few experimental schools have been established on the basis of that flexibility.

An empirical confirmation of the principal's powers can also be found in the fact that when new schools are opened they are rarely opened as independent schools, but rather as branches of an existing school until they pass through a "colonial" period of development and are deemed by the local department of education to be entitled to autonomy.

What of the parents in this situation? Education is one area in which there is a great deal of parental concern—if not a concern to participate in the educational process as such, certainly a desire to determine what that process will involve for their children and under what conditions the children will be learning. The parental role is manifested in several ways. First, the school day in Israel is relatively short and the amount of homework great on the assumption that parents will undertake some of the responsibilities of teaching. Many, if not most, parents also believe they have the right to choose the school in which their children will be enrolled. They clearly have that right with regard to which subsystem they choose. This is entirely a parental decision, by law. Indeed, the law mandates local authorities to provide school facilities appropriate to parental demand and the Ministry of Education must provide the funds for constructing the necessary buildings and providing the teachers once the local authority certifies that the requisite school population exists for one subsystem or another.

Beyond that, while the principle of school districts and the neighborhood school does exist in some measure, it is frequently honored in the breach, with parents seeking a school for their children which they find desirable from their point of view. Moreover, the local authorities themselves are prone to circumventing their own districting if they believe that a school will be better balanced by mixing students from different neighborhoods, provided that they can persuade the parents accordingly.

Parents additionally involve themselves in the schools on a regular basis through parent associations, parent-teacher conferences, and a sheer feeling of freedom to visit the school and speak to their child's principal or teacher at will. In fact, there is close contact between principals, teachers, and parents in most cases because of the educational tradition that has been fostered in Israel. By law, parents also have the right to alter up to 25% of the curriculum of their school in consultation with the principal. Here, too, the same reality has prevailed as in the case of the principal's powers in this regard. Outside the kibbutzim and certain experimental schools, there is not 25% of the curriculum that any but a handful of parents would want to alter.

Kindergartens, prekindergarten education, and high schools are the direct responsibility of the local authorities, albeit with financial and technical assistance from the Ministry of Education. This is because the state has determined that it cannot yet afford to provide free compulsory education through the high school years for all children or absorb pre-first-grade education. Thus, tuition is charged for these programs—in the case of pre-first-grade education a very nominal sum, in the case of high schools a rather large sum. While the Ministry of Education provides funds that enable this tuition to be waived, the present situation still limits the number of people continuing their education beyond 10th grade.

The same subsystems prevail on the high school level. The major unifying force is the system of matriculation examinations required by the Ministry of Education and prepared, administered, and graded by Ministry of Education personnel on a uniform basis throughout the country. In the days before the spread of university education, the high schools possessed the status of colleges; their teachers were often great scholars, and the principals were lords. Every school had a very definite point of view, orientation, and even methodology.

Many were sponsored by agencies of the Jewish people rather than by indigenous public institutions, a situation which continues to prevail. Thus, technical schools are mainly in the province of ORT (Organization for Rehabilitation and Training, a worldwide Jewish body), and even today most high schools are built through special donations from Diaspora Jews channeled through the Jewish Agency.

WELFARE

Welfare is formally a cooperative state-local service in which the localities operate welfare programs funded in whole or in part by the Ministry of Welfare. The operation of welfare programs is similar to that of grant-in-aid programs in other countries. The localities have responsibility for determining who is eligible under criteria promulgated by the Ministry of Welfare. They create the packages of welfare benefits to be given to any individual or family on the basis of the various programs provided by law, and they furnish the social services needed to assist the family in rehabilitation or adjustment to its condition.

As in other countries, the effectiveness, efficiency, and responsiveness of welfare programs are regularly attacked, both in the press and in studies. At the same time, it is clear that Israel does not suffer from the masses of permanent welfare cases that have come to exist in the United States. Nevertheless, as the population in Israel sorts out, the lowest stratum is moving in that direction, and there are already cases on record of several generations of welfare clients from the same family. Israeli practice, on the other hand, has been to prevent the use of welfare to sustain the lower levels of the population, preferring instead to provide "make-work" for the people of marginal employment ability so that they can retain their self-respect and remain off the welfare rolls. This is coupled with a wide variety of social benefits provided through the Institution for Social Insurance, the Israeli equivalent of the U.S. Social Security Administration, out of its central office in Jerusalem.

LOCAL FUNCTIONS

There are a number of functions that are purely local, among them garbage collection, libraries, and parks. With the exception of the first, which tends to be provided at a uniformly high level by localities around the country, these vary from locality to locality, depending upon the degree of interest on the part of the governing officials and relevant pressure groups in securing proper facilities. Israel's local park systems are relatively undeveloped, partly because this kind of amenity requires a sophisticated population for its support. Much the same is true for libraries. In both cases, capital expenditures and operating funds are mobilized largely from outside the community, the former from overseas contributors and the latter from the state government via the Ministry of Education.

Here the scale question seems to be one of minimum rather than maximum size; that is to say, there are local authorities which are too small to maintain effective libraries or to be concerned about developing parks. A 15,000 to 20,000 population may be the cutoff point here, although there are striking examples of relatively poor development towns of 5,000 population providing library services equal in quality to those in the biggest cities, because someone has made an effort to do so and has managed to mobilize internal and external support.

SUMMARY AND CONCLUSIONS

Israel is a country which offers a wide variety of possibilities for testing questions of scale and governmental organization with regard to their effectiveness, efficiency, and responsiveness. As a small country divided into a relatively large number of local jurisdictions, it reflects many of the advantages of small to moderate scale. Its history of development of multiple jurisdictions to handle specialized local governmental tasks, despite a formal tradition which denigrates that approach, suggests how natural that approach may be regardless of other factors militating against it. At the same time, Israel's special political culture and a strong tendency toward bureaucratization influences the impact of scale. These bear investigation if we are to learn how the influence of scale fits into the influence of other components of a political order. To date, little research has been done on these questions in Israel, although there is much social research available about various aspects of Israeli institutions, particularly the kibbutz which, given the proper tools, can be analyzed for some benefit.

Most of the Jews who came to the country came after the state was established and not as pioneers. In general, they had very low expectations regarding government services and even lower expectations regarding their ability to participate in or even influence the shape of those government services. The expectations of the Arabs, on both counts, were even lower. At the same time, many of the Jews were ambivalent in that they saw the new state as a messianic achievement and hence expected its government to solve personal problems of housing and employment in a very paternalistic way.

As the population acquired an understanding of democratic government, their demands intensified, passing, in some cases, from passivity to almost unrestrained insistence on having their way. With this escalation of demands came an escalation of complaints about the way in which services were delivered. Individuals would seek to influence those responsible for service delivery in specific cases affecting them, relying heavily on personal contacts to do so, but saw no general role for themselves as participants in the political process. This is now slowly changing, as more and more native born Israelis reflect the socialization process of the school system and what we have come to associate with middle-class values in the political sphere.

By and large, there has been no systematic effort on the part of the public or spokesmen for the public to articulate and aggregate public preferences into collective choices that become the expression of demand. With the exception of a few areas of immediate concern that are tacitly understood as such, matters in Israel have not much passed the grumbling stage.

By the same token, it would be hard to say that there is a conscious effort to organize the production and delivery of services in response to such demands. That is obvious enough since the demands themselves are hardly felt to exist. Much of what does exist in this regard is a result not of internal pressures in Israel but of leaders being cognizant of the trends in the Western world in this direction and seeking to find echoes of them or perhaps to anticipate such echoes within their country.

With that initial understanding, it is possible to identify two major sources with regard to the articulation and aggregation of demands. There are protest groups that have emerged from among the disadvantaged members of Israeli society, that is to say, those immigrants from Oriental countries and their children who have been left behind in the general upward mobility of the population. They have made substantial claims, particularly in the areas of education and housing, upon all the authorities of the state on the grounds that they are suffering from discrimination and lack of equal opportunity.

The form of those demands has followed traditional lines of protest rather than systematic efforts to transform the present situation.

The other group consists of members of the academic community whose business it is to study urban problems and make recommendations for their resolution. These people have been tempted to follow conventional Western European and American thinking on the subject, accepting the management-oriented reformism of the 20th century West. However, virtually all of those who have been in responsible positions have, whether for reasons of political prudence or intellectual skepticism, refrained from pursuing those ideas as far as they have been pursued in the West. Thus, in a major report for the reorganization of urban services in the Dan region, the task force refrained from recommending conventional metropolitan consolidation and instead recommended a multiple-purpose metropolitan authority, limited in scope, that would not eliminate existing municipal government.

In addition, there is a small but significant group of academic dissenters, who share many of the views of the "public choice" school with regard to urban public service delivery. That is to say, they are impressed by the empirical evidence that the theses of the management model regarding size, fragmentation, overlap, and the like are inadequate if not just plain wrong.[9]

Whether these forces are sufficient to overcome bureaucratic inertia and the natural preferences of a people who have grown accustomed to a hierarchical system is an open question. What is clear is that the political culture of Israel acts as a strong bulwark against changes in the present system—a system that balances formally hierarchical and centralized institutional structures against a myriad of implicitly contractual arrangements, with all the bargaining and negotiation that accompany such arrangements and actually inform the system. Perhaps as the Israeli political culture takes on a more consistent and harmonious character, this will prove to be dysfunctional, and one or the other aspect will undergo serious modification.

NOTES

1. The data reported in this article are as yet unpublished and are obtainable in the files of the Institute for Local Government. For sources on Israel local government, see Yehezkel, 1975.

2. Lands purchased by the Jewish National Fund are deemed to be the permanent

possession of the entire Jewish people for whom the JNF serves as trustee. They cannot be alienated through sale but only through long-term lease to those who work them or who develop them for useful purposes. Virtually all Jewish agricultural settlements, the kibbutzim (communes) and moshavim (smallholders' cooperatives) are located on JNF land which they hold by lease. The terms of the leases include social provisions with regard to proper land use and require the observance of the Sabbath in matters connected with the property on the part of the leaseholder (many of the latter provisions have proved legally unenforceable but retain some moral authority).

3. Some kibbutzim have developed the servicing of groups and individuals from the Diaspora into an "industry." Kibbutz Kfar Blum, for example, sponsors a high school for English-speaking Jews. Many kibbutzim sponsor three-month intensive language programs *(ulpanim)* which combine half a day's Hebrew study with half a day's work on the land, for which the kibbutzim are compensated by the Jewish Agency.

4. There is a great deal of misunderstanding regarding the Law of Return. Israel has immigration laws similar to those of other Western countries, with permits issued upon application and naturalization following in due course. However, since Israel is considered the state of the Jewish people, Jews enter almost as if they were engaging in interstate migration in the American manner. It should be noted that similar laws hold true in other countries with regard to those considered nationals even if born outside their borders.

5. It must be emphasized that the separations which result are by choice and not by law. While Arabs may go to any school in Israel, they prefer to maintain their own schools, in which the principal language of instruction is Arabic rather than Hebrew and the curriculum reflects Arabic culture and either Muslim or Christian religious beliefs and practices.

6. It would not be incorrect to estimate that as many as one-third of all Israelis hold the religious law of their respective communities in higher regard than the law of the state, including a small group of Jews (perhaps several hundred) who reject state law altogether.

7. The resemblance between the Israeli system and consociational arrangements is equally apt (Lijphart, 1974).

8. Length considerations prevent a full discussion of the impact of Israeli frontier experiences and patterns of development on the delivery of local public services. However, several points should be noted. First, the rural-land frontier stage which began in the late 19th century and continued into the early 20th remains an important aspect of Israeli development—with population settlement and the formation of new communities in rural land areas continuing to occur. Second, the urban industrial frontier stage in Israel was shaped by a number of factors: the urban industrial frontier stage began relatively late (between the First and Second World Wars, but with most development activity occurring after 1948); the country was small and lacked navigable waterways (which meant that roads rather than waterways, railroads, or internal air networks became the principal means of transportation and the automobile, the principal vehicle); and the country lacked raw materials and large indigenous or adjacent markets (which meant that the kind of heavy industry upon which the urban-industrial frontier was based in the West was, for the most part, beyond Israel's capabilities). With few exceptions (Haifa and the small cities located adjacent to it, for example), one would be hard put to find an Israeli city whose growth was due to the urban-industrial frontier per se.

9. Some, like David Pines, have done theoretical work to demonstrate that such things as "urban sprawl" are economically valuable aids rather than hindrances to metropolitan development. Others, like Ephraim Torgovnik, have translated articles into Hebrew presenting the public choice hypothesis (Torgovnik, 1976). Still others, like Moshe Hazani, are doing empirical research into problems of size and scale from a sociological perspective, whose results are strengthening the public choice theses (Hazani, 1976). Hazani, for example, has shown the negative impact of the unilateral intervention of the Ministry of Housing in one Israeli community that theoretically is too small to plan its own housing (in Israel this is a function of the central government in any case) but which undoubtedly

would have done a better job than the bureaucratized intervention of a ministry whose plans at that time were primarily the products of theoretical considerations and current architectural styles without significant reference to the site or community involved.

REFERENCES

ALDER, M. (1960). "Local government in Israel." In Public administration in Israel and abroad. Jerusalem: Israel Institute of Public Administration.

BALDWIN, E. (1972). Differentiation and cooperation in an American veteran Moshav. Manchester: Manchester University Press.

COHEN, E. (1970). The city in Zionist ideology. Jerusalem: Hebrew University, Institute of Urban and Regional Studies.

CRIDEN, Y., and GELB, S. (1974). The kibbutz experience. New York: Herzl.

EISENSTADT, S.N. (1967). Israel society. London: Weidenfeld and Nicholson.

ELAZAR, D.J. (1970). Cities of the prairie: The metropolitan frontier and American politics. New York: Basic Books.

––– (1971a). Israel from ideological to territorial democracy. New York: General Learning Press.

––– (1971b). "Kinship and consent in the Jewish community: Patterns of continuity in the Jewish communal life." Tradition, 14(4).

––– (1973a). "The local elections: Sharpening the trend toward territorial democracy." In A. Arian (ed.). The elections in Israel. Jerusalem: Jerusalem Academic Press.

––– (1973b). The metropolitan frontier: A perspective on change in American society. Morristown, N.J.: General Learning Press.

FEIN, L.J. (1967). Politics in Israel. Boston: Little, Brown.

GUTMANN, E. (1958). "The development of local government in Palestine." Unpublished Ph.D. dissertation, Columbia University.

HARTZ, L. (1964). Founding of new societies. New York: Harcourt, Brace and World.

HAZANI, M. (1976). The anatomy of a local community. Ramat Gan: Bar-Ilan University Institute of Local Government. (Hebrew)

JAFFE, M. (1966). "The authority of the rabbinate in Israel." In Public administration in Israel and abroad. Jerusalem: Israel Institute of Public Administration.

LAQUEUR, W.Z. (1972). History of Zionism. New York: Holt, Rinehart and Winston.

LIJPHART, A. (1974). "Consociational democracy." In K.D. McKae (ed.), Consociational democracy: Political accommodation in segmented societies. Toronto: McClelland and Stewart.

LIPSET, S.M. (1963). First new nation. New York: Basic Books.

MELJON, A. (1966). Local government in Israel. Tel Aviv: Israel Union of Local Authorities.

ORNI, E., and EFRAT, E. (1974). Geography of Israel. Jerusalem: Israel Universities Press.

OSTROM, V. (1971). The political theory of a compound republic: An analytical essay on the Federalist papers. Blacksburg: Virginia Polytechnic Institute Center for the Study of Public Choice.

TORGOVNIK, E. (1976). "Local policy determinants in a centrist system." Unpublished manuscript, Tel Aviv University.

TURNER, F.J. (1962). Frontier in American history. New York: Holt, Rinehart and Winston.

WEISS, S. (1969). "Results of local elections." In A. Arian (ed.), The elections in Israel. Jerusalem: Jerusalem Academic Press.

YEHEZKEL, M. (1975). Bibliography of local government in Israel. Ramat Gan: Bar-Ilan University Institute of Local Government.

3

Cooperative Arrangements for the Delivery of Public Services in the German Democratic Republic

KENNETH HANF

☐ IN RECENT YEARS a number of scholars have "turned to the concept of bureaucracy to understand the nature of the Soviet system" (Hough, 1973:135). According to Hough, "much of the popularity of the bureaucratic model, in its various guises" seems to lie in the fact that it has "permitted Western scholars to retain many of the traditional notions of totalitarian control" even as they acknowledge the obvious changes that have occurred in the post-Stalin period (p. 145).

With such models, students of advanced socialist systems have stressed the continuing hierarchical nature of the political system through which society is mobilized and organized in accordance with the party's will. Thus A. Meyer writes that in all bureaucratic systems "the system remains unified under one command, coordinated by one universally binding set of goals, guided and controlled by one central hierarchy" (Meyer, 1965:468). However, in light of what we know about the dynamics of large-scale organizations, such an understanding of bureaucracy underestimates the problematic nature of hierarchical control and neglects the important role that bureaucrats can play in the formulation as well as the execution of policy. It also ignores the importance of interactions between a given organization and different elements in its environment.

An adequate model of the dynamics of policy making and planning in these countries must recognize that the organizational arrangements used by governments that seek to deal with an increasingly significant range of problems are shaped by the highly dynamic, complex, and interdependent environment associated with social and technological change. The policy process is confronted with a set of tasks where both the problems and their solutions cut across the boundaries of separate authorities and functional jurisdictions. At the same time, the extensive institutional differentiation —both functionally and territorially—of the politico-administrative system makes it unlikely that any individual unit or any single level of government will possess the necessary jurisdiction or resources to encompass the relevant dimensions of most problems or programs.

The ability of individual decision units to achieve their own objectives depends not only on their own choices and actions but also on those of others; actions at any one level of decision making are influenced by relationships that exist between levels and across functional boundaries. Consequently, a crucial institutional problem confronting all modern industrial states is the construction of arrangements through which the actions of separate, yet interrelated, decision units can be coordinated in carrying out the comprehensive programs of action required for steering socioeconomic development under conditions of a post-industrial society.[1]

Under these circumstances, an analysis that places too much emphasis on central decision makers and hierarchical relations between decision levels may overlook the significance of coordinative activities at the local level. Such an approach fails to appreciate the increasingly significant "pieces of political action" that occur in relationships between central and local units of government. To conclude from the undeniable concentration of decision making power in the hands of central decision makers that local units are nothing more than instrumental administrative subdivisions of central government is to overlook the impact of the reciprocal relations between central and local actors on the formulation and implementation of policy. A more complete picture of the relative "weight" of different levels of government in their various and complex interactions thus requires an examination of patterns of cooperation and coordination among the different elements of the structurally and functionally differentiated system of state control.

This essay describes some of the salient features of the system of state action through which a wide range of governmental services, intended to achieve the "steady improvement of the working and living conditions" of the citizen, are delivered at the local level in the German Democratic Republic (GDR). Its focus is on the various activities that contribute to the provision of these goods and services and the relationships between the different actors involved. It takes as its point of departure the contention that effective problem solving requires the mobilization and coordination of the activities of a number of separate institutional actors in increasingly complex sets of decision situations. Hierarchy is, to be sure, the dominant element in most decision situations in socialist systems. The central party leadership and, in its name, the central organs of the state are, both in theory and in practice, preeminent. Nevertheless, in an increasingly complex setting it is doubtful that the central leadership will be able to integrate activities behind general system goals by means of hierarchical commands alone. Rather than prejudging with the choice of model the ability of central leadership to direct members of organizations and to coordinate the activities of different levels, we must place the problem of integration and coordination at the center of analysis. In this way, we can examine the mechanisms and strategies that are employed by the political leadership "for integrating an increasingly complex, heterogenous and structured environment, even as it maintains a centrally directed hierarchical one-party system" (Breslauer, 1976:67).

This essay focuses, therefore, on the response of the political leadership in the GDR to the problems of complexity and interdependence confronted in the attempt to mount a centrally directed, unified program of societal development. As part of a larger study of the delivery of social services in the GDR, the materials presented here are limited to a general description of some of the basic features of the institutional arrangements that have been developed over the last few years. Because of the difficulty in the GDR of observing governmental operations, or even interviewing local governmental officials, the picture drawn highlights the understandings of the leadership rather than the actual functioning of the system constructed. The description focuses primarily on the formal legal steps taken to realize the objectives laid out by reformers. It seldom penetrates to the actual behavior of the various actors involved. Nevertheless, insofar as the constraints upon, as well as opportunities for action by, different individuals and groups in this system are defined by formal rules and the assignment of "resources" to positions, such a description can *suggest* some of the structural factors that may affect actual behavior.[2]

DEVELOPED SOCIALIST SOCIETY AND DEMOCRATIC CENTRALISM IN THE GERMAN DEMOCRATIC REPUBLIC

THE PRIMARY TASKS OF SOCIAL POLICY

A basic thesis of Marxist-Leninist political theory is that the state is a product of society on a particular level of development. From this insight it is held to follow that a political order cannot be understood apart from an analysis of the mode of economic production and the resultant social relations that form its material base. Accordingly, it is argued that an examination of the features of the socialist state must begin with an analysis of the economic and social conditions characteristic of a given stage in the unfolding of socialist society. Insofar as the socialist state will be subject to "the lawfulness of social development," it will undergo continuous change in the course of the further construction of socialism.

At the same time, however, this socialist state is not viewed merely as a reflection of its social and economic substructure. On the contrary, it is considered to be the main instrument through which the working class is actively to transform society and create the conditions for the ultimate realization of communism. If this political instrument is to meet the demands made upon it in the construction of this new social order, it is pointed out that its structures and operations must be adapted to the socioeconomic context in which it is embedded and which it is supposed to shape. Thus, with each step taken toward the achievement of communism, the tasks and procedures of the socialist state will necessarily change; its social base will expand and be strengthened; and the conditions shaping the nature and effectiveness of its activities will be modified.

With the conclusion of the construction of the foundations of socialist society in the early 1960s, the GDR is considered to have entered a relatively extended period of dynamic and comprehensive social and economic development. In the course of this development, it is expected that the assumed advantages and driving forces of socialism will come to complete expression in all areas of social life and that the conditions for the full unfolding of the "economic laws of socialism," as the basis for rational and effective economic policy, will be fulfilled. This period of societal development, described as the "shaping of the developed socialist society," is thus defined in terms of both the "objective processes and tendencies" at work in society and the goals and tasks guiding the movement of socialist society toward communism.

This period is seen as a time of change that requires the bringing of "all sides and all areas of social life to full development" (Kalweit, 1976:3). Successful realization of policy objectives during this phase will, it is felt, depend upon mastery of the complex interrelationships and interdependencies that exist among the various social and economic processes at work. Policy and planning must thus be directed not only at "the successful development of individual areas of social life" but also at the realization of "the correct interrelationships among them" (Kalweit, 1976:3).

The draft of the new party program presented for public discussion prior to the Ninth Party Congress in May 1976 lays out the tasks to be performed during the present phase of socialist construction. It begins with the Marxist-Leninist notion of the two phases of a unified communist social formation. During the "first phase of the communist society" the major tasks are consolidation of the political power of the working class and full development of the socialist mode of production. It is from an analysis of the differences that exist between the socialist and communist phases, and the various measures required to overcome them, that the strategic orientation for the activity of the state and all other organizations in the GDR phase of development is derived.

First Secretary Eric Honecker, in his report of the Central Committee of the Socialist Unity Party (SED) to the Eighth Party Congress in 1971, sketched the primary tasks to be fulfilled in constructing mature socialism (Honecker, 1972:38). The long-range social and economic program is aimed at raising the material and cultural standard of living of the population by means of a high rate of economic development. It is from this broad statement of the objectives of state policy (reaffirmed at the recent party congress) that the concrete economic and social policies for the present phase of socialist development are to be derived.

It is repeatedly pointed out in the literature of the GDR that the pursuit of these goals is an "objective necessity" and their realization a "real possibility" because the country now possesses a productive potential for creating the material wealth on which the continued improvement of the general standard of living depends. It is stressed that only the continued growth of the economy can provide the resources for what has been described as the "most comprehensive social program until now in the history of the GDR" (Kalweit, 1976:3). In this sense, the principal set of tasks remains further development of the economic system.

Although economic tasks will continue to occupy a central position, the increased productivity of the socialist economy in the GDR has made it possible to devote more attention and resources to social programs. The improvement of the lot of the citizen has, it is argued, always been a goal of socialism. At the center of the new primary task, however, is the desire to create a more immediate relationship between economic production and the satisfaction of the needs of the citizens. Not only is the productive process to take these needs as a point of departure but conditions are to be established to ensure that increases in the level of production are felt more immediately in the form of improvements in the quality of individual and collective life.

Within the framework of these broad principles the political leadership has laid out a comprehensive social program through which the genral welfare of the population is to be improved. This general policy commitment has since been operationalized through numerous measures aimed at improving housing conditions, increasing the supply of consumer goods and services, continuing developments in the area of health and social services, and improving the general and vocational training for young people and adults. All of this is to be accompanied by an increase in the real income of the population through both wage increases and public expenditures. As an indication of the seriousness and scope of these programs, reference is made to a reported increase of 35% in state expenditures in the 1970-1974 period for such things as education, health and social services, social insurance, and public subsidies for stable prices and rents.

THE SYSTEM OF GOVERNANCE

The GDR is a unitary state. Its territory is divided into 15 districts, 218 counties, and 8,850 cities and towns of various sizes. Each of the four levels has a popularly elected assembly, an executive council chosen by the assembly, and a number of administrative departments (under the direction of members of the council) through which the administration of government programs is carried out. All government units below the central level are referred to as "local organs of state power."

The supreme state organ at each level is the popularly elected assembly. In keeping with the tradition of assembly regimes, these bodies unite all governmental powers in themselves. There is no system of separated powers distributed among constitutionally independent institutions.

In an important sense, the basic structure of the state in the GDR is the system of popular assemblies that runs from the People's Chamber at the center down through the succeeding levels of government. However, the state apparatus, in the narrower sense of the word, consists of the system of executive councils with their administrative departments. These administrative activities may be separated into those administered through the hierarchy of executive councils at the different territorial levels (general administration) and those administered through field agencies of the central ministries. To the latter belong such activities as police and security, military affairs, customs, state attorneys, foreign affairs, and foreign trade. Post, communications, and railroads are public enterprises with their own administrative structures.

A central concern of the socialist state—and one that accounts for the great profusion of state agencies—is the direction of the economy. The economic apparatus is headed by the various industrial ministries with the chain of command running from these central ministries through an intermediate level of economic direction (the Associations of Socialist Enterprises) down to the individual production units in different parts of the country. This hierarchy is responsible for effecting central direction of economic production.

In addition, approximately 25% of industrial production is carried on in plants directly under the control of local executive councils. These are concerned primarily with consumer goods and services, foodstuffs, and auxiliary production for larger enterprises. The direct supervision of agricultural production is also centered at the county and communal levels. Local state organs are not supposed to interfere in the production activities of the centrally directed economic units in their territory. However, they do have a major responsibility for coordinating economic activities with regard to the use of territorial resources and other preconditions of production.

In addition to this economic hierarchy, the state works closely with various social organizations at different territorial levels. Such organizations as trade unions, youth groups, cultural associations, and the National Front (the federation in which these organizations and the political parties are joined) play important roles in the realization of different aspects of state policy. For example, the social insurance program is administered by the Free German Trade Union Federation; it is also responsible for the plant safety inspection program. At the county and communal level, local state organs collaborate closely with the National Front in mobilizing local resources for various community projects.

Surrounding these institutions is the Socialist Unity Party (SED), the supreme political force in the country. The party makes basic decisions regarding the ends and means of developmental policy. Its decisions, which define the general interest of society at a given time, have the force of binding directives for all state officials. Membership of high state officials in the central directive organs of the party join the party and state hierarchies. This same arrangement is found at local levels as well. The finely spun network of party organizations —in the form of party groups within the state agencies and the party apparatus itself—makes it possible for the party to penetrate the state agencies at many different levels.

Although the state is intended to be the instrument of the party, a division of labor does exist between state and party. The party is ordinarily supposed to restrict itself to decisions on fundamental matters of societal development. The concrete realization of this policy is the responsibility of state agencies.[3]

DEMOCRATIC CENTRALISM AND UNIFIED STATE ACTION

The structures which have been developed in the GDR (and in other socialist systems) reflect a particular view as to the kind of relationship that should exist between society and the party. The goal of the party is to promote the balanced and proportional development of society and to allow, within this framework, a full development of the individual parts. It is argued that the development of subsystems must contribute to an effective development of the whole. In the GDR the consequences of this holistic perspective are preeminence of the center as an articulator of the general social interest and rejection of the notion that local interests are legitimate apart from their contribution to general social development. This basic assumption regarding the preeminence of the national or general social interest is supported by the concentration of political power in the hands of the SED.

Closely related to the preeminence claimed for the decisions of central decision makers is the belief that measures required at any given time to realize long-run developmental goals can be "objectively" known. This claim reflects and reinforces the dominance of the SED as the ultimate source of political power and wisdom. A further institutional implication is the dominant role played by central state decision makers in operationalizing party policies into concrete programs. These central party and state organs lay out fundamental policy goals and programmatic measures. This framework is intended to serve as the binding context for developments and activities at each successive level of government. Through it, local organs of the government are provided with the direction and support necessary to place their activities within the broader developmental context.[4]

The organizational principle on which this structuring of the state apparatus rests is democratic centralism. Through the application of this principle, a hierarchically organized state apparatus has been constructed which runs from the People's Chamber down to the town assemblies and from the Council of Ministers to the executive councils on the communal level. In this way, a dual chain of command is established through which decisions and information are transmitted from top to bottom and the behavior of subordinate organs is continually monitored.

These vertical chains of command are complemented by a horizontal one. As we have noted, the executive councils at each level are elected by their respective popular assemblies. Directors of local administrative departments are appointed by local executive councils, with participation by assemblies at the same level and agencies at the next higher level. Together, these vertical and horizontal lines of responsibility create an overall system of double subordination. In this way, the system of state direction and planning is supposed to ensure unified action at all levels in pursuit of central policy objectives while allowing implementation of these policies to vary in light of the differential needs and possibilities of local territories.

At present, the further development of democratic centralism is to be guided by a growing awareness of the limits to hierarchical direction from the center.[5] Although democratic centralism (and the political realities it reflects) does not allow for decentralization of power to autonomous local units, it is compatible with various degrees of deconcentration in the interests of greater flexibility and effectiveness in pursuit of general policy objectives. Consequently, moves have been taken to provide opportunities for more meaningful local action in mobilizing territorial resources.[6]

DELIVERY OF STATE SERVICES AT THE LOCAL LEVEL

LOCAL GOVERNMENT AND SOCIAL POLICY

Commentators in the GDR argue that with qualitative changes in the national economy and with the requirements for realization of the proclaimed comprehensive social program, the scope of state activity will necessarily expand (Weichelt, 1972:27). If, however, the state is to meet the demands being placed upon it, the state apparatus must also become more flexible and effective in its operations and more sensitive to the needs and views of the citizens, whose energies it must mobilize in pursuit of policy objectives. Success of state action in directing the development of socialist society is viewed as depending on the adaptation of its structure, procedures, and activities to the changing nature of its tasks and the conditions under which they are performed.

The new responsibilities of the state cannot be met solely by strengthening the capabilities of central state organs. According to commentators in the GDR, the stronger and more effective that central state planning becomes, the more crucial that the planning and directive functions of local state organs will be for preparation of central plans and their implementation. In this sense, the "further development of the power of the socialist state is inseparably connected with the perfection of direction and planning by the local organs" (Schüssler, 1973:41). It follows that one of the

> logical steps in the application of democratic centralism lies in the continuous expansion of the role of the local organs in order to consolidate further the power of the working class, to extend without interruption its mass base, and to shape more effectively the exercise of [this] power. [Schüssler, 1973:41]

Local levels of government in the GDR already play a significant role in performing those tasks through which the objectives of social policy are realized. In particular, local units exercise considerable control over the production of consumer goods and services and are also directly responsible for supplying a wide range of government service functions in their respective territories. Local assemblies and their executive councils are, for example, responsible for

- approximately two-thirds of the building supplies and construction industries;

- 30% of the transportation services, in particular commuter, public, and school transportation;

- 80% of total retail sales;

- 70% of the public funds distributed by the state for cultural and social services—especially for schools and for social and health care; and

- the main part of the production of consumer goods, particularly food stuffs and light industry (Böhm, 1972:53).

In this regard, the local levels—especially the cities and towns—represent crucial points of interface between the citizen and the system for delivering goods and services.[7] These contacts are of importance for the relation between the citizen and state in a more general sense. Most citizens "experience" their state in the person of the officials with whom they have contact at local governmental offices. Their evaluations of the performance of the state are greatly influenced by these "face-to-face contacts" as well as the quality of services received. These experiences obviously color the feelings that citizens have, vis-à-vis the state; and, thus, can influence their long-run identification with the political order.[8]

Local assemblies are also important mechanisms for the mobilization of social energies behind national development goals. Development programs of the scope to be pursued in constructing developed socialist society make increasingly higher demands upon the individuals who, in their various capacities, are called upon to carry out the tasks. If such programs are to be successful, citizens must commit their full energies to the fulfillment of the responsibilities that they have in meeting plan and program goals. To this end, it is hoped that wider popular involvement (under the guidance of higher state authorities) in planning and implementation of these programs will contribute to a larger measure of agreement and, ultimately, harmony between personal interests and the requirements for continuing territorial and societal socioeconomic development.

In the local assemblies all "social forces" in the area are supposed to take part in shaping the development of the territory and in integrating these processes with state policy as a whole. Thus, as the active involvement of the citizens becomes more crucial, popular assemblies and their organs will become increasingly important in providing opportunities for participation in deliberations over territorial development, as channels for information exchange and feedback and for mobilizing the activity of citizens in pursuit of these objectives.

LOCAL GOVERNMENT IN THE SYSTEM OF STATE POWER

Relations between local and central units of government depend on whether "the function of local governments is solely or primarily to foster local interests or to serve national interests at the local level" (Leemans, 1969:19). Clearly, if nationally determined programs and objectives are considered preeminent, not only will "a certain structure of local government and a certain relationship between local and central government be required," but it will be necessary to shape "local government units in such a way so that their size and managerial capacity guarantee the level of achievement and service demanded by the center" (Leemans, 1969:19).

In the GDR, central direction and the effective transmission of central decisions throughout the state apparatus are viewed as preconditions for the extension of local responsibility. This means that local organs of the state are intended to function within a larger intergovernmental system through which the general policy of the state is realized. According to this model, the fundamental responsibility of these local organs is to be the effective implementation of a centrally determined uniform state policy.[9]

Nevertheless, there exists a certain amount of flexibility in the principle of democratic centralism. On the one hand, the preeminence of central decisions means that decision premises, information resources, and so forth flow from the top down. On the other, the differential conditions under which these policies are to be implemented must be acknowledged and adjustments made to local differences in social structure, needs, and resources.

Accordingly, at each level of the hierarchy, local activities are to be derived from central policy objectives and concretized for different levels and conditions. This means that local policy is supposed to make its specific, self-responsible contribution to the realization of these goals differentiated according to the particular local area. It is asserted that these responsibilities cannot be adequately realized if equated with a mere execution of matters commissioned or predetermined by central authorities. On the contrary, local decision makers are supposed to be in a position to mobilize their own resources within policy parameters and program goals set by higher decision makers. The parameters set by higher levels are thus supposed to be broad enough to allow adequate room for such initiative and responsible action. At the same time, they must be specific enough to guarantee that local activities will remain on the generally desired path of development.

In concrete terms this means that, while local organs are to decide "self-responsibly" on all matters of fundamental importance for their territories, they do so on the "unshakable basis" of party decisions, laws and resolutions of the People's Chamber, decisions of the Council of Ministers and of higher popular assemblies and executive councils (Schulze et al., 1972:15). Such an arrangement is supposed to ensure that "central state direction and planning will be firmly combined with local initiative and that the point of departure for local decisions . . . will be the . . . policy of the working class and its party" (Schulze et al., 1972:15). It is in this sense that these territorial units are described as "organs of the unified socialist state."

THE SYSTEM OF LOCAL ORGANS OF STATE

In order to understand the dynamics of decision making under democratic centralism in the GDR it is also necessary to consider an equally important structural feature of this system: its high degree of differentiation. It is argued that uniformity of action, as the means for achieving a concentrated impact on the realization of central policy objectives, is not to mean that everywhere, without distinction, the same thing is done. On the contrary, a certain division of labor will of necessity exist, where "in the solution of specific problems and the fulfillment of particular responsibilities, each organ and each functionary works uniformly on the solution of the tasks of the socialist state as a whole" (Weichelt, 1972:98). While these activities are to be based on the objectives contained in central plans and decisions, it is pointed out that the goal of uniform state action cannot be sought at the expense of reduced responsibility at the different levels of state direction. The system of central direction rests upon a complex division of labor between levels of state direction and between different organs of state power.

An unambiguous delineation of the respective tasks and functions of each level of state action is viewed as an indispensable precondition for raising the authority of local units and strengthening their role in the system of state direction and planning. By avoiding duplication and overlap in the responsibilities of local organs, it is expected that officials will be given a clear orientation for concentrating their attention and energies on those matters for which they are held accountable. Such a focus, it is argued, is the prerequisite for mobilizing their creative initiative and forward-looking responsibility for development of the respective territories.

The formal demarcation of this system of differentiated responsibility is provided in the 1973 Law of Local Assemblies and Their Organs. According to this law, each level of local government has been given a clearly defined set of functions together with the formal rights and powers held to be necessary for their performance. In allocating decision authority, an attempt was made to match, as far as possible, the dimensions of the problem and the requirements for its solution with the resources and expertise of the different levels. Proceeding in this way, responsibility for a given task has been located as far down in the hierarchy as is consistent with its efficient performance and effective control by superior decision makers.[10]

Such a distribution of tasks and responsibilities is intended to reflect the differing scales of governmental functions and the resulting need to locate responsibility for particular tasks at a level which encompasses the relevant dimensions of the problem. Basic to the division of labor across the levels of local government in the GDR is the role played by superior organs in "disaggregating" general policy responsibility into sets of tasks for self-responsible implementation through subsequent levels. Districts and counties, through their decisions, define the parameters within which the particular level below them is expected to organize its own activities. In particular, these superior levels are to work out the long-range perspective on territorial development and set points of emphasis for the activity for all social, political, and economic organizations within a territory for a particular planning period. Superior organs are also supposed to play an important role in assisting subordinate organs of state power in planning and carrying out their respective activities in connection with the achievement of central policy objectives.

In addition to these functions, administrative tasks are also performed at the district and county levels. These activities are associated with the direct provision of services, particular aspects of development, and the production of economic units under their control. Districts and counties are concerned, thus, with the more "far-reaching" matters. Cities and towns on the other hand, deal with those affairs involving direct and continuing interactions between the state and the citizens. These local bodies have been defined as self-responsible communities in which citizens work, live, and shape the conditions of their social life. As such they are supposed to create the conditions for satisfying the material, cultural, and social needs of their members. To this end, cities and towns have the primary responsibility for those activities that affect the immediate, daily lives of citizens.

COMPLEXITY, COORDINATION, AND COOPERATION

ORGANIZATION FOR COMPLEX ACTION

We have noted that the activities of local organs of state power are directed toward a comprehensive and balanced socioeconomic development of each territorial unit on the basis of centrally determined parameters. In pursuit of this objective, local decision makers must act within a highly dynamic, highly interdependent environment. In recognition of this fact, commentators in the GDR have repeatedly pointed out that the "realization of the primary task in all areas requires a new approach to the solution of these tasks" (Schulze et al., 1972:11). That is, the primary task requires that local assemblies and their executive councils approach their responsibilities in a "complex manner."

These commentators stress that "with the shaping of developed socialist society a qualitatively new phase has begun in which, with regard to state directive activity, important new conditions are given" (Klein and Oertel, 1975:1011). The present period of socialist construction has been described as one of "extraordinary intensification of the interrelations between the forces of production, the relations of production and the social superstructure" (Reinhold, 1973:72). This manifests itself in the increasingly dynamic and complex character of the processes of social change. As a result, "the tasks to be solved have become more varied and complicated and are, in particular, of an increasingly complex nature" (Klein and Oertel, 1975:1011). In order to deal with the growing complexity of social and economic problems and with the high degree of interdependence among them, programs aimed at the proportional and balanced development of society as a whole have become an even more important concern of state activity. The formulation and realization of such programs requires the mastery of complex interrelationships among the various social processes. As a result, the "conscious shaping of the dialectical unity of all sides of socialist society" is to be a central focus of state directing and planning activities at all levels (Rost, 1973:1649).

At the heart of recent institutional reforms has been the attempt to improve the capacity of each level of government to handle the complex tasks confronted in shaping social and economic development. Despite all efforts to match organizational structure to clearly delineated tasks, the fact remains that problems do not always coincide with the problem-solving capacities of any particular decision unit. This is the case for problems that cut across functional jurisdiction at a given level; it holds equally where problems transcend the boundaries of individual territories. The key problem in this regard is structuring the decision system so that plans of action appropriate to the dimensions of the problem can be formulated and the contributions of a variety of decision units to their implementation can be concerted. The general organizational answer to this problem has been to turn to various forms of coordination.[11]

It is possible to distinguish a vertical and a horizontal dimension along which such coordinative action is sought. The system of hierarchical relations is designed to coordinate the activities of different levels of state direction by providing the general goals and decision parameters by which actions at each level are integrated and related to one another. But while a directive hierarchy may integrate activities vertically, it may also lead to the creation of narrow pillars of responsibility along functional or programmatic lines. In itself vertical control does not assure broad, comprehensive action.

In this sense a complex approach to problem solving means that the actions of particular units must be integrated horizontally as well as vertically; different decision makers must coordinate their actions with those of others, in a way that is relevant to a particular program or problem. Specific actions must also be considered and taken in the broader context of the general development of the program and the territory and in terms of the long-range objectives pursued.

There are a number of devices which are intended to supplement the vertical line of direction in this regard. For example, collective decision making in the executive councils at all levels is viewed as an important means for countering narrow departmental interests and perspectives. Here it is expected that the level of complexity will be raised by bringing together different interests within the context of general policy considerations. Planning commissions are also supposed to balance different strands of territorial development while preparing integrated economic plans. In addition, the planning methodology applied by all units of government lays out procedures through which coordinated program plans are to be developed at each level of state direction. Viewed in this context, subordination of the functional administrative departments (to local assemblies and to superior administrative agencies and the state) is also intended to integrate program interests toward the social development of the territory. It is by means of such arrangements that the continuing attempt is made to direct the activities of the individual institutional actors toward the realization of general policy objectives.

TERRITORIAL COORDINATION:
ECONOMIC ENTERPRISES AND SOCIAL SERVICES

A major objective of state policy is the rational utilization of all resources for the balanced social development of the different territories as the precondition for both improving the working and living conditions of citizens and developing optimal territorial conditions of production. In pursuit of these goals, local organs of the state and economic units are mutually interdependent. The correct shaping of these relationships is considered to be a crucial factor affecting both economic growth and the quality of life in a given territory.

The balanced development of the national economy, in turn, depends not only upon harmonious development of different sectors of the economy but upon the proper intermeshing of sector development with territorial requirements and possibilities. Overall responsibility for balancing and integrating sectoral and territorial considerations rests with the Council of Ministers and its central organs. However, management of these interrelationships on a continuing basis is the immediate responsibility of local assemblies and their executive councils. The effective performance of economic units in a given area depends upon the availability of various factors of production, many of which are supplied by the territorial unit in which the enterprise is located. Local organs of the state have important functions in supplying these territorial conditions of production—such as labor, raw materials, technical infrastructure, and so forth—and for coordinating the various demands made upon the territory by different sectors of the economy and their respective units of production.

Economic units not only make demands upon territorial resources but also make important contributions to the overall social development of the territories in which they operate. All plants are responsible for the quality of working and living conditions of the workers they employ. Thus, it is not unusual—especially in the case of larger enterprises—for plants to supply nurseries for workers' children and laundries for workers' families. Plant health services run from immediate first-aid facilities to elaborate "policlinics," clinics with highly specialized medical staffs.

The quality of life of workers is, of course, not the result of plant conditions alone. It is also influenced by the social conditions in the broader community in which they live. Likewise, measures taken within the plants have an impact on social development beyond the factory gate. These interrelations are seen clearly when one compares city and economic unit expenditures for social programs. The city of Freiburg, for example, with a population of 50,000 expended 3.2 million Marks for its cultural and social programs in 1968. During the same period the mining combine located in the city (and employing 6,500 persons) spent something more than 3 million Marks from its funds on related activities. Within the city there were about 60 other plants with active programs of this kind. In the district of Cottbus, plants of the soft coal industry operate plant cafeterias dispensing 34,000 meals daily, 10 nurseries and 20 kindergartens, 300 places in vacation homes, 15,700 factory owned housing units, 21 cultural houses, 9 ambulatory medical facilities, and 9 policlinics (Zienert, 1968:1597).

Thus, over and above the production of goods and services, economic enterprises are themselves essential components of the system for delivering many social services. Coordination of actions and resources of plants and local units of government is expected to open up new opportunities for improving the living conditions of the general community. In this sense, economic enterprises and local organs of state power in each territory are seen as two sides of a single process of social development with joint responsibility for the quality of life in their particular territory.

In light of increasing interdependencies between economic enterprises and governments in towns and cities, it is argued that a close and effective collaboration between them has become indispensable. The Law on Local Assemblies defines the general principles governing relations between economic units and local organs of the state. With this law, the powers of local governments with regard to coordination of all activities in their respective territories were specified for the first time. At the same time, administrative codes regulating the status and responsibilities of economic units obligate these units to work closely with local state organs in achieving a balanced social, cultural, and economic development of territories. While enterprises are responsible to their superiors in the economic hierarchy for fulfillment of production plans, they are also responsible to local state organs for meeting their obligation to promote the social development of the community.

At a minimum, such coordination involves the exchange of information with regard to ongoing and planned programs. Among other things, this means that economic units are required to present their plans for improving working and living conditions to the local authorities in their municipality. At the same time, they are expected to make proposals for joint measures in these areas.

Another crucial element in the coordination of the social functions of enterprises and local organs of the state has been the use of contracts. With such contracts, cities and towns can "buy" particular services from the economic units. Even more important, however, is the use of such agreements for mobilizing funds and materials for joint construction and utilization of facilities, ranging from bus depots to community culture houses and swimming pools.

On the basis of such contracts, for example, eight Berlin enterprises and the executive council of one of the city's districts recently agreed to collaborate on the expansion of the children's camp facilities of one of the large factories for joint use by workers' families from smaller factories which were parties to the agreement. Another large plant agreed to supply approximately 1,240 meals daily to two schools and about 20 middle-size factories in the area. In addition, free lunches are to be supplied for pensioners. Agreement was also reached on cooperation in renovating a number of apartments for workers. Citizens of another city district in Berlin have profited from joint use of the health facilities of a large enterprise. The policlinic at this factory serves around 10,000 workers from other small and middle-size factories, approximately 50% of the district's population, as well as its own work force (*Neues Deutschland*, 1976:8).

Various regulations spell out the conditions under which local units—as organs of state power—can obligate enterprises to take certain actions. At the same time it has been stressed by commentators that such "obligating" should not become the dominant type of relationship between enterprises and territorial organs of state power. These reciprocal relations can be shaped more effectively, it is argued, through the exchange of information and common deliberations and planning. Action should be taken in close consultation and agreement with the economic units involved, thus making unilateral "binding obligations" the exception (Schubert and Gerlach, 1973:44).

In line with this approach, cooperation based on continuous interaction between enterprises and local government in all phases of the planning process is viewed as an essential precondition for territorial development. It must begin with the projection of long-range developments in different areas, move to the synchronization of projects, and lead to agreements on specific joint endeavors. In this way, particular actions are to be integrated into the comprehensive development of the territory as a whole.

SOCIALIST COOPERATION AMONG LOCAL COMMUNITIES

Further development of the national economy is assured to require increasing concentration and specialization of productive forces. This will mean a greater concentration of social life in particular cities and towns and an expansion of the economic and social service functions from these centers to surrounding areas. Social relations between citizens in different towns and cities will thus become closer and the interdependencies between the local communities more pronounced. These developments set the conditions under which programs for improving the working and living conditions of citizens must be planned and carried out. Despite collaboration with economic units in their territory, all but the largest cities are not in a position, alone, to set up and operate the various programs and facilities needed to care for their citizens. Moreover, the development of an effective territorial structure itself requires that more attention be paid to the functions performed by cities and towns for their surrounding areas. There is also a need to plan these social developments in terms of units larger than the individual municipality. Within this context it has been assumed that questions of community development can only be dealt with through integrated policy making and action on major issues for an entire area.

There are many ways of dealing with the questions raised by the frequent incongruence between problem areas and territorial structures. In some countries this situation has led to direct intervention by higher government authorities in local affairs and the assumption, by regional or national governments, of tasks previously assigned to local levels. Efforts to create more appropriate local units have led also to moves for annexation, amalgamation, or the redrawing of administrative boundaries according to selected functional criteria.

In the GDR it is continually stressed that socioeconomic development programs contain an important political element directed toward the active participation of citizens in the steering of this development. Therefore, in redesigning the local governmental structure for the delivery of these social services, they are reluctant to create governmental units that would make state organs and decision processes more remote from the citizen and make direct involvement more difficult. In addition to the normative considerations of citizen participation, the argument for a direct "local presence" is also made in the name of efficiency. An arrangement that keeps responsibility as low as possible, it is argued, makes government more flexible and responsive to local needs and allows for important informational inputs from the populace. Most important, it is expected that such structures are more effective in mobilizing directly the energies and initiatives of the people called upon to carry out or support the measures to be implemented.

Thus, the design rule in the GDR for matching problem scale to organizational unit is: bigger but not more remote. The structures developed must provide for continued direct participation by, and responsibility to, the various municipal units involved. In applying this rule, the GDR has turned to arrangements for cooperation and collaboration among cities and towns (and between cities and towns and economic enterprises as discussed above) as a way of creating large enough service areas and production units to take advantage of anticipated economies of scale. In this way, they hope to make certain activities economically feasible and technically efficient at the local level. The result of this policy is a growing network of direct cooperative relationships among cities and towns throughout the country.

At present the two most developed forms of collaboration are the joint authority *(Zweckverband)* and the association of communities *(Gemeindeverband)*. The joint authority is

> a form of continuing socialist collaboration by cities and towns—on the basis of concurrent resolutions of the popular assemblies of these communities—which has, as its objective, the improvement of the working and living conditions and the more rational utilization of funds, through the joint use of forces and resources in dealing with a particular activity or problem. [Gläss et al., 1972:53]

Such authorities have been widely used, for example, to supply street maintenance services, housing and public facilities, service centers, or recreation facilities and services.

The activities of joint authorities are limited to specific tasks or problems. The association of communities is a more comprehensive form of collaboration across a number of program areas. As presently conceived, the association is intended to serve as a framework for formation of larger territorial units. Eventually, communities participating in these associations will become subunits of the new local governmental units. It is also possible for a town, which is a member of an association of communities, to contract with other towns and cities if the contract is in the interests of the association and contributes to the improvement of working and living conditions in the particular community. The association itself can also be a member of joint authorities to concentrate its resources for improvements in living conditions of its members.

Towns and cities in the GDR are not only "local" organs in charge of directing and planning the social development of their territories; they are also organs of the unified system of state power. Thus, it is not surprising that the planning and setting up of joint decision-making arrangements occurs within parameters set by development plans for the more inclusive territorial units of which the cities and towns are a part.

Formally, all decisions regarding formation of cooperative arrangements lie with the popular assemblies of the participating communes. Entry into a joint authority, for example, is left to the voluntary choice of those communities involved, and each participant retains its autonomy as a territorial unit responsible for social development in its own jurisdiction. In practice, however, decisions regarding the need for such collaboration, are not left solely to local authorities. Such cooperation takes place with the active guidance and assistance of superior party and state decision makers. Indeed, the initial stimulus for such cooperation is likely to emanate from higher party and state officials in consultation with the mayors of these communities (the chairmen of the respective executive councils) and the directors of the major economic enterprises. County officials, in particular, may be actively involved in judging the "ripeness" of conditions for collaboration, in assisting and guiding work by local organs on the longer-range development plans, and in providing direct inputs into decisions and plans made.

The involvement of superior organs is designed to ensure that joint projects by local organs fit into the larger territorial picture. By ensuring that questions of social development will be considered from the point of view of the county and the district, these superior officials are supposed to prevent mistakes in the disposition of communal resources and energies. At the same time, county officials are not supposed to "interfere" with the self-responsible activity of the local communities. The higher decision-making level is not to make the determinations itself; but rather to create the conditions necessary for effective collaboration. Counties need not give their formal approval with regard to most joint operations. Nor can county officials require that cooperation be undertaken or order the reorganization of an ongoing collaborative arrangement. On the other hand, because of its impact on future developments in the territorial governmental structure, an association of communities does require confirmation by the county, after prior approval by district authorities.

Close cooperation between local organs of state power and their superiors at the county and district levels means, however, that final approval is a "mere" formality. The process of coordination between hierarchical levels during the formative phase will guarantee general agreement among the parties involved. Indeed, the fact that a joint decision-making arrangement has been set up is, itself, evidence that the necessary consensus among different officials has been achieved.

In an important sense, these collaborative arrangements are designed to be "mobilization nodules" within the general system of state direction. As such they are intended not only to concentrate financial and material resources but to tap the energies of the individuals on whom successful goal achievement depends. These human agents include not only state officials with direct responsibilities for meeting program objectives but citizens as well. It is repeatedly stressed that goals set for joint operations can only be met if *all* citizens are involved. To be sure, superior officials can prod and cajole; they can set limits and steer the efforts of local participants; they can retain the last word on what is to be done. However, if they are not able to create an appreciation of the need for collaboration on the part of those most directly affected, the goals of the collaboration will not be achieved. The emphasis on the voluntary commitment of local communities to joint responsibilities thus rests on a recognition of the importance of the "subjective element" in the direction and planning of social development.

Creation of the necessary preconditions for effective cooperative relations is viewed as a gradual process which moves from elementary forms of consultation to increasingly more complex forms of collaboration. The ability of officials and citizens of participating communities to recognize the "objective" need for such cooperation depends on whether or not previous cooperative experiences have promoted a sense of responsibility that goes beyond the boundaries of their own town. In order to promote an understanding of the need for collaboration, intensive ideological and political activity is an essential part of the formation of joint decision-making systems in the GDR.

CONCLUSION

The design of governmental institutions in the GDR has been influenced by a concern for uniform implementation of the policy of the socialist state, and for guaranteeing that territorial development reflects the centrally defined needs of society as a whole. At the same time, this unified system of state power is designed to incorporate a high degree of functional and structural specialization. As a result, the unity of state policy will itself be highly complex and differentiated, to be achieved only by putting together the necessary "program package" from among the individual units involved. Although the initial coordinating impulse is supposed to come from above, an equally important dimension of such a system will be collaboration among state decision makers at the local level and with social and economic organizations in their territories. A variety of such mechanisms for jointly mobilizing resources have been developed, within the framework of a basically hierarchical system.

There is nothing new in noting that any highly differentiated decision system will require a variety of forms of cooperation and coordination. It is also true that, in the past, the activities of individual state organs in the GDR had to be coordinated with one another. However, this integration now occurs under different social conditions and in the context of new political dynamics. The necessary coordination for such conditions cannot be achieved solely by means of hierarchical commands emanating from above. As a result central decision makers today must confront the de facto influence accruing to local organs by virtue of their "functional" importance in coordinating territorial development and improving the working and living conditions of citizens in an advanced socialist society. These institutional arrangements will produce further developments in the social and economic processes they are intended to direct. In this sense it can be argued that the same factors that have "brought about" the need for a system balancing central direction and self-responsible local action will provide an important, albeit limited, guarantee for maintaining the system they have "called forth." If this system is to function in the intended way, the political leadership would seem to be forced to recognize that it cannot promote more extensive delegation of responsibility on the one hand and fail to provide the necessary decisional power and financial and material resources for meeting these responsibilities on the other. Of course, these leaders could choose to forego effective performance of the economic and social policy system in the interest of maximizing or protecting other values (i.e., the power of the party and its central leaders). However, in the face of demands for larger quantities of consumer goods and services, the extent to which this option is realistically available is an open question. Consequently, continued local self-responsibility rests on the expected function it fulfills in improving the working and living conditions of citizens and in stabilizing the regime.

The system of governance in the GDR, while incorporating competing organizational principles, clearly is tilted toward its hierarchical dimension and control from above. There is no denying that the decision system is characterized by two basic realities: the priority of the general social interest as authoritatively interpreted by the SED, and the predominance of central decision makers in determining the content of this general interest and the measures necessary for its realization. These features are also reinforced by ideology and the distribution of power as well as the inertia of bureaucratic organization. There remains a strong tendency to restrict local initiative in a system where hierarchy is so prominent. This tendency is supported by the inclination of higher officials to clear up uncertainty and confusion with directives and interventions from above. Nor are local officials always willing to take the responsibility that the system offers.

Other factors are at work which pull the system in a different direction: the legitimacy of "democratic participation," the organizational requirements of effective performance, the de facto "weight" of subordinate levels and units in decision making, and the generally problematic nature of hierarchical control in complex organizations. There are many situations where a central organ of the party or state is in a position to "coordinate" the relevant actions through binding decisions. There are equally important nonhierarchical situations in which it is not possible to command coordination but where it can be achieved only through cooperation. In general, it is stressed by GDR commentators that, even where it is possible to issue unilaterally binding commands, a more effective basis for productive relationships lies in some form of cooperation.

Even within the constraints of democratic centralism there are signs of an opening of the decision-making processes to participation by subordinate units. Institutional channels are being provided through which individual and group interests may enter into central determinations of the general social interest. Further research and more unrestricted access to information will be necessary to establish the extent to which these possibilities for influencing central direction and planning from the "bottom up" have resulted in effective articulation of the interests of noncentral participants in policy formulation and have developed self-responsible mechanisms for coordinative actions in a complex manner among local enterprises and organs of the state.

If we are to enhance our understanding of these interdependencies and measures to deal with them, we need an analytical perspective that focuses on linkages and interrelationships among different levels and decision units. It is important that future research take, as its point of departure, an understanding of democratic centralism as a network of organizations with a dominant but, by no means, all-powerful center.

NOTES

1. For a systematic treatment of many of the questions raised in connection with this problem (and illustrative case studies), see Scharpf et al. (1976).

2. It is clear that a behavioral analysis is required before we can evaluate the actual impact of these measures on the policy process in the GDR. The literature in the GDR on these developments tends to be heavily normative, making it difficult to determine what is, in fact, going on. Moreover, deviations from the desired pattern of behavior are quite often attributed to idiosyncratic human factors which, it is argued, can be corrected by more training and ideological enlightenment; seldom is such "undesirable" behavior seen as stemming from the organizational arrangements themselves and the interactions between and within these structures. Yet there is, a priori, no reason to doubt that human behavior in large-scale organizations—whether in socialist or capitalist societies—will display important similarities. On the basis of this assumption a study is now being prepared to test certain propositions regarding organizational behavior reflecting and influencing the dynamics of policy making in socialist countries. This study will probe a number of factors likely to affect such behavior and describe patterns of interaction between different agencies

and levels of administration. It is part of a larger research project being carried out by David Porter and the author. It seeks to extend some of the insights suggested in Thompson (1967) and Williamson (1970) and developed further in the theory of resource mobilization presented by Porter (1973).

3. It seems hardly necessary to note that such a division of labor raises a host of questions concerning the possibility and the desirability of making such a distinction between policy and administration or between major policy decisions and implementation. This same problem occurs in delineating the responsibilities of different levels of the state apparatus and the respective roles of the assemblies, their commissions, the executive councils, and the administrative agencies at each level. As is often the case, GDR commentators resolve such tensions with normative statements about the preferred relationship, relegating problematic aspects of actual behavior to individual deviations from the norm. The above-mentioned behavioral analysis will examine the actual role of the "administrator" in both policy formulation and implementation. The wealth of experience in the United States and Western Europe, and the literature analyzing it, will serve as a point of departure for our analysis of socialist countries.

4. For an excellent discussion of some of the general questions raised in designing arrangements that provide central direction while allowing for the flexibility held to be necessary for effective operations, see Porter (1976) and the literature cited there.

5. Obviously, the problems raised with regard to organizing and managing large-scale organizations—or intergovernmental systems—under conditions of bounded rationality are not limited to capitalist societies alone. The design of central-local systems in socialist countries must also come to terms with the conditions discussed by Simon, Lindblom, or Wildavsky.

6. The tendency of reformers and commentators in the GDR to stress the normative aspects of formal arrangements obscures possible—and more-than-likely—conflicts between different interest and power groups within the state apparatus. There is, however, an interesting ambivalence in the attitude of the GDR commentators on this issue. On the one hand, they deny the possibility (and certainly the legitimacy) of such conflicts in the correctly functioning socialist society and view deviations as signs of persisting remnants of capitalist attitudes and behavior or as an indication of an underdeveloped socialist consciousness. On the other hand, they seem willing to admit the continued existence of nonantagonistic contradictions and conflicts in society (related to the unevenness of societal development as well as to the need to set priorities with regard to the production and consumption of scarce resources) and to recognize and utilize the "material interestedness" of individual actors in lining them up behind general social goals. For an interesting discussion of this last point see Mampel (1975). "Management theory," as it is presently developing in Eastern European countries, contains treatments of problems of motivation and leadership styles not dissimilar to those found in Western literature.

7. In many important ways the GDR—as well as other advanced socialist societies—differs little in this respect from nations of Western Europe and the United States. Certainly a comparison of the activities performed by local or subnational units of government would show few significant differences between these countries. A preliminary report on a project comparing the range of governmental services and the factors accounting for variations in the patterns of these activities at the subnational level in four Eastern European countries is provided by Welsh (1976).

8. Porter (1975) discusses the growing importance of the interface between the citizen-consumer and governmental agencies at the local level and makes some interesting observations regarding the kinds of factors that should be taken into account in organizing these points of contact more effectively.

9. Once again it should be noted that the model underlying recent institutional reforms in the GDR rests heavily on a number of unrealistic assumptions regarding the politics-administration dichotomy and the passive subordination of local interest and power

groups to higher-level decision makers and interests. No doubt both reformers and superior decision makers are well aware of the potential and actual deflection of the central will by the actions and the inertia of lower decision units. Their answer to this problem of "bureaucratic power" is the somewhat contradictory strategy of tightening control and specifying plans and programs in more detail from above while, at the same time, seeking to encourage more "self-responsible" activity on the part of local units and to incorporate these subordinates—where it is felt appropriate—in decision making by the higher units. This strategy, of course, reflects the basic organizational problem of squaring the circle of central control and local discretion. The point is, however, that the correct balance cannot be achieved by definition of the desired roles alone. At the same time, the organizational alternatives available for dealing with this problem are "given" by the fundamental features of the socio-political system in question. In this regard, the role of the Communist party in mediating and structuring relationships among levels and individual actors is extremely important in the GDR.

10. With this law the consequences of a number of years of practical experience and analysis have been drawn and codified. In attempting to deal with the problems of matching function and area, the GDR confronts questions not unlike those wrestled with by students and reformers of metropolitan government and by those concerned with shaping an effective system of intergovernmental relations in the U.S.

11. For a discussion of some of the basic issues raised here, as well as a particular set of answers informed by the perspective of the "new political economy," see Bish (1971) and Bish and Ostrom (1973).

REFERENCES

BISH, R.L. (1971). The public economy of metropolitan areas. Chicago: Markham.

BISH, R.L., and OSTROM, V. (1973). Understanding urban government. Washington, D.C.: American Enterprise Institute.

BOHM, S. (1972). "Finanzpolitik im Interesse des werktätigen Volkes." In Der Fünfjahresplan dient dem Wohle der Arbeiterklasse und des ganzen Volkes. Berlin: Staatsrat der DDR.

BRESLAUER, G. (1976). "The soviet system and the future." Problems of Communism, (March-April):66-71.

GLASS, K., et al. (1972). Probleme der Entwicklung der sozialistischen Gemeinschaftsarbeit zwischen Städten und Gemeinder der DDR. Potsdam-Babelsberg: Akademie für Staats-und Rechtswissenschaft der DDR.

HONECKER, E. (1972). Bericht des Zentralkomitees an den VIII (Parteitag). Berlin: Dietz Verlag.

HOUGH, J.F. (1973). "The bureaucratic model and the nature of the Soviet system." Journal of Comparative Administration, (August):134-167.

KALWEIT, W. (1976). "Die Einheit der beiden Phasen kommunistischer Gesellschaftsformation." Neues Deutschland, (January 30):8.

KLEIN, D., and OERTEL, H. (1975). "Bedeutung der staatlichen Leitungstätigkeit für die Ausnutzung der Vorzüge des Sozialismus." Staat und Recht, (7):1702-1714.

LEEMANS, A.L. (1969). Changing patterns of local government. The Hague: International Union of Local Authorities.

MAMPEL, S. (1976). "Die Funktion des Rechts bei der Bewaltigun von interessengegen Satze." Deutschland Archiv. Sonderheft: 69-90.

MEYER, A.G. (1965). "The Soviet political system. New York: Random House.

Neues Deutschland (1976). "Sozialistische Gemeinschaftsarbeit führt zu besseren Arbeits- und Lebensbedingungen." (February 26):8.

PORTER, D.O. (1973). The politics of budgeting federal aid: Resource mobilization by local school districts (Sage Professional Papers in Administrative and Policy Studies 03-003). Beverly Hills, Calif.: Sage.

——— (1975). "Responsiveness to citizen consumers in a federal system." Publius, 5(fall):51-77.

——— (1976). Adapting the "profit center" concept to government administration. Berlin: International Institute of Management.

REINHOLD, O., et al. (1973). Die entwickelte sozialistische Gesellschaft. Berlin: Dietz Verlag.

ROST, R. (1973). "Zu einigen Problemen der weiteren Vervollkommnung der staatlichen Leitung nach dem VIII: Parteitag der SED." Staat und Recht, 10/11.

SCHARPF, F.W., REISSERT, B., and SCHNABEL, F. (1976). Politikverflechtung: Theorie und Empirie des kooperativen Föderalismus in der Bundesrepublik. Kronberg/Ts: Scriptor Verlag.

SCHUBERT, K., and GERLACH, G. (1973). Ortliche Staatsorgane und nicht unterstellte Betriebe. Berlin: Staatsverlag der DDR.

SCHULZE, G., et al. (1972). Richtig Entscheiden—Wirksam Kontrollieren. Berlin: Staatsverlag der DDR.

SCHUSSLER, G. (1973). "Der demokratische Zentralismus als Grundprinzip der staatlichen Leitung und Planung." Staat und Recht, 1.

THOMPSON, J.D. (1967). Organizations in action. New York: McGraw-Hill.

WEICHELT, W. (1972). Der sozialistische Staat. Berlin: Staatsverlag der DDR.

WELSH, W. (1976). "Toward explanations for within-system policy variations in Eastern Europe: A preliminary research note." Unpublished manuscript.

WILLIAMSON, O.E. (1970). Corporate control and business behavior. Englewood Cliffs, N.J.: Prentice-Hall.

ZIENERT, H. (1968). "Ständige Verbesserung der Arbeits- und Lebensbedingungen der Werktätigen—Gemeinsame aufgabe sozialistischer Betriebe, Städte und Gemeinden." Staat und Recht, 12.

4

The Structure and Performance of Urban Service Systems in Italy

PHILIP SABETTI

□ A CONCERN FOR THE PERFORMANCE of urban service systems has been a persistent theme in Italian public life. In recent years the increasing deterioration and inadequacy of public services in most urban centers have produced new manifestations of this concern among citizens and public officials.

Urban service systems seem unable to cope with elementary aspects of everyday living. Relatively simple services, such as street paving and lighting, garbage collection, and water supply, are not properly discharged. Standards of public health remain unenforced. Air and noise pollution and traffic congestion have become exceptionally serious. Empty buildings in densely populated areas are becoming common, as are shantytowns and illegal housing developments—both devoid of essential public services. Crime against persons and property is increasing, as is the employment of private guards and private security forces. Italian cities that were once an essential part of European Grand Tours are now shunned by tourists. Both

AUTHOR'S NOTE: *The formulation of this paper has benefited significantly from the comments and suggestions of Frances Pennell Bish, Ezio Cappadocia, Judith Johnson, Vincent Ostrom, and Mario Stoppino. The essay draws in part upon earlier work supported by Canada Council/*Conseil des Arts du Canada *and the Workshop in Political Theory and Policy Analysis at Indiana University, help which I wish to acknowledge.*

urban and rural centers seem strangled by the heavy hand of bureaucracy.

Response to the eroding quality of urban life has included a number of changes in governmental policies and structures aimed at improving public service performance and at bringing government closer to the citizens. Since 1950, professional training of police officers and other bureaucratic personnel has increased; the expansion of public employment rolls has been coupled with substantial salary increases for state officials; unprecedented expenditures for public works have taken place. These new policies have been accompanied by institutional changes. During the 1960s, "little city halls" and district councils were established in large urban centers as part of communal decentralization. In 1970 regional government, already operative on the islands and in the Alpine areas, was extended to the rest of the peninsula as part of the decentralization of the national bureaucracy. Most of these programmatic and institutional changes coincided with the Opening to the Left, hailed as the new era in Italian public life.

By the early 1970s, it became apparent that this new era in Italian public life had not markedly improved conditions of life. "Pathologies" in the delivery of urban public services continue to exist. The disparate efforts of citizens and public officials to deal with these problems are based on a common concern: Why the persistent failure in the performance of urban service systems?

The answer to this question is dependent upon a knowledge of the characteristic results of urban service systems. What are the characteristic results which flow from the political process? If we can arrive at some initial assessment of these results and pursue an examination of why urban service systems produce them, then we may be in a better position to answer the question of why the persistent failure in the performance of urban service systems in Italy.

This paper represents a preliminary attempt to deal with the structure and performance of urban public service delivery systems in Italy. The two dominant factors influencing the delivery of urban services in Italy are the highly centralized state administrative and decision-making structure and the system of electoral laws which encourages the active participation of national parties in the organization of local party activities. These two factors are considered as a part of the basic design of Italian political institutions. Next, we turn to an analysis of their consequences for urban public service delivery using the cities of Naples and Bologna as examples. In the last section, recent reforms at the city and national levels designed to "decentralize" decision making are described and the short- and long-run consequences of these reforms are discussed.

THE DESIGN OF INSTITUTIONAL ARRANGEMENTS FOR PROCURING AND DELIVERING PUBLIC SERVICES AT THE LOCAL LEVEL

Contemporary institutional arrangements for the procurement and delivery of public services in Italy rest on design criteria used to fashion a united Italy in the 1860s. This nation state was organized as a single center of authority with an exclusive monopoly of the ultimate use of force in the organization of society. The monopoly over the supply of public goods and services was accompanied by a single overarching system of public administration with local elected officials and professionally trained personnel hierarchically ordered and subject to direction by heads of departments at the center of government (Fried, 1963:79-93; Petracchi, 1962).

The Fascist experience strengthened this monocentric state. Following the collapse of the Fascist regime, the essential elements of the pre-Fascist parliamentary regime were reintroduced and a constitutional charter created new institutions, such as a constitutional court and regional governments, aimed at limiting central government authority. Secondary laws introduced during the Fascist era remained, however, and Fascist reforms, ranging from communal institutions to the judicial system, became permanent fixtures of republican Italy—impeding the full implementation of the constitutional charter.

THE MONOCENTRIC STATE AS A CONSCIOUS CHOICE

The monocentric solution to the question of Italian self-government in the 1860s was a conscious response by the artificers of unification to several perplexing problems. The new nation-state was forged through military victories. Previous rulers of Tuscany and Lombardy had allowed a large measure of self-government in internal affairs; the same conditions had existed in Sardinia and Sicily until the Napoleonic wars. This tradition of self-government gave rise to a strong attachment to community affairs as well as an extraordinarily diverse set of social institutions, cultures and languages. To permit inhabitants of the peninsula to continue their self-governing capabilities—as Lombard and Sicilian publicists urged (Cattaneo, 1945; Ferrara, 1949, 1965)—would have, it was assumed, annulled the military victories and destroyed the very foundations of the state (Cavour, 1949, vol. 1, p. 305).

The forced creation of unity through administrative measures under a common parliament, backed by a national army, appeared to Piedmontese leaders to be the only alternative. This forced creation of unity sought to forge the diverse communities of peoples into one strong and great nation, to insure a uniform provision of public services, and to remove once and for all the specter of foreign interventions in Italian affairs (Cavour, 1949, vol. 1; Pizzorno, 1971:90).

Sicilian analysts reached different conclusions about the consequences likely to flow from this forced unity. They suggested that the operation of the monocentric state would produce the opposite and *not* the expected results (Ferrara, 1949). The principal artificer of Italian unification, Cavour, rejected this analysis. He averred instead that his

> theory of the State [did] not imply either the tyranny of the capital over the rest of Italy or the creation of a bureaucratic caste that would subjugate all other bodies and would thus transform the position of the Government into an artificial center of an empire which the traditions and habits of Italians and Italy's geographic configuration would always be against. [Cavour, 1949, vol. 4, p. 220]

The history of the first 60 years of the Italian nation-state is essentially the history of how successive central governments —whether of the left or the right—struggled to maintain a monopoly over the supply of public services. From the perspective of central government officials, citizen dissatisfaction with this state of affairs reflected a lack of patriotism; attempts by local officials to be responsive to citizen preferences or local interests were perverse manifestations of responsiveness; local efforts to secure home rule or to assert an inherent right of self-government were expressions of parochial values, and mutual aid societies were mafias or criminal associations. As Einaudi later noted (1954:52), the artificers of unification "believed that they were establishing liberty and democracy when they were forging the instruments of dictatorship."

The Second World War revealed the fragility of the forced creation of unity. The reopening of the question of Italian self-government after World War II was accompanied by the emergence of political groupings which claimed to represent the nation. The strategic position acquired by the Demochristian and Communist-Socialist groups allowed their leaders to become the major artificers of "the second rebirth of Italy."

These new artificers were confronted with the same perplexing problems that Piedmontese leaders had faced a century earlier. Changes in the authority and powers of the instrumentalities of the central government—as some analysts were urging (Einaudi, 1954; Levi, 1963:248-254)—might have given support to localized groups intent on asserting an inherent right of self-government and demolished the work of the fathers of *Risorgimento* or unification. Moreover, the left and Demochristian leaders, who participated in the emerging postwar conflict between the two international superpowers, viewed the instrumentalities of the monocentric state as an opportunity to apply the principles of Marxism or Christian Democracy and to advance or retard the cause of "Eastern" or "Western" civilization.

There appeared once again to be no alternative to the reiteration of the monocentric solution. During the postwar years, the central government thus retained authority over (1) the design of institutional arrangements at the local level, (2) the choice of public services provided by these instrumentalities, (3) the exercise of governmental prerogatives in relation to the supply of public services, and (4) the monitoring of the performance of public service systems.

COMMUNAL AND PROVINCIAL GOVERNMENTS

The Italian peninsula and islands are divided into 8,000 communes or municipalities or varying sizes. Each commune has the same governmental structure and authority originally granted by the central government. This general uniformity or "civic equality" derives from the fact that communes, and not individuals, were the basic units for the fashioning of the Italian state.

In 1934, a general law on local government was promulgated to make communal and provincial institutions "more *efficient* and *responsive* to the *demands* of the central authority" (Steiner, 1939:315; italics added). This was achieved in two ways: consolidation and annexation. In 1920 there were 9,000 communes and a population of over 30 million people; by 1934 the number of communes had been reduced to 7,000 and the population had increased to 40 million.

The trend toward consolidation or annexation was reversed following the Second World War, and approximately 800 communes were either reconstituted or established anew. These changes, however, left unaltered the 1934 general law that had strengthened the authority of the central government over the organization and conduct of communal affairs. Indeed, the only element of discontinuity in the structure of local government between the Fascist and the republican regimes is the election of local public officials (Rotelli, 1973:155). In the postwar period each commune has served as (1) a local office for the vast bureaucracy of the central government, (2) the local administrative unit for the ministry of the interior, and (3) an expression of "local government" or "local autonomy" in the sense that elected communal officials are responsible for the production and delivery of communal services.

The structure of authority relationships inherent in the conduct of national affairs is reflected in the conduct of communal or municipal affairs. The position of the mayor and members of the executive committee *(giunta)* vis-à-vis the communal council and the communal bureaucracy is exactly the same as that of the prime minister and members of his "council" vis-à-vis the national parliament and the national bureaucracy. Large cities such as Milan and Naples have 15 or more communal departments ranging from sanitation to urban planning, but each department is organized in relation to a member of the executive council charged with general responsibility for the procurement and delivery of that service. The same principle applies to the operation of municipal enterprises. Since the turn of the century, municipal enterprises have produced and delivered services ranging from public transportation to water and gas under the direction of administrative boards appointed by and responsible to the mayor and the executive committee.

The choice of which services are supplied by the central government or the communes depends upon the importance of each service *for* the central government. Services such as education, public health, public security, and fire protection are supplied by the central government organized in terms of ministries or state technical services. Each technical service has a local production unit engaged in supplying a specific good or service that is jointly used by local residents. Each production unit is organized upon monocentric principles. The size of cities affects only the number of technical personnel: the larger the city, the larger the group of technical personnel.

For example, a typical village is likely to have a medical health officer from the public health administration, a forest warden heading the forestry services from the ministry of agriculture, a commandant of *carabinieri* in charge of the local garrison of the state police officers, and a "didactic director" in charge of elementary school instruction. With some differences, similar arrangements prevail on a larger scale in cities. Fire protection service, for example, is generally supplied only to city residents. Inhabitants of small communes depend on the fire protection service of the provincial fire brigade. For every city and provincial capital the maintenance of public order or crime detection is performed by two parallel state police forces: the *carabinieri* from the Ministry of Armed Forces and the public security corps *(polizia)* from the Ministry of the Interior. At the same time, in an effort to maintain the parallel system of police throughout the entire territory, mayors of small communes without public security officials act as officers of the public security corps vis-à-vis the commandant of the local *carabinieri* garrison. In this capacity, mayors of small communes grant, for example, stay permits to foreigners wishing to reside in their communes.

Until recently, provinces served in part as intermediate field and control levels for local public service systems. By World War I, there were 75 provinces. During Fascism the consolidation of communes was accompanied by the creation of another 19 provinces to improve the monitoring of the performance of communal governments and state technical services. After World War II, while a number of old communes were reconstituted, the number of provinces established during Fascist rule remained; the reduction from 94 to 92 provinces resulted from a loss of territory to Yugoslavia. These territorial divisions remained for the convenience of the national bureaucracy. The number of communes attached to each province varies considerably, however. Some northern provinces such as Turin still have as many as 300 communes. Each provincial capital reproduces in a microcosm the state apparatus that exists at the national level.

During Fascist rule, the provincial prefect exercised general control over all instrumentalities of government operating within the confines of each province. After the war, this general control over all provincial instrumentalities of the central government was withdrawn. The prefect remains, however, as the major representative of the central government at the provincial level and the chief field agent for the Minister of the Interior. In this capacity, each prefect exercises general control over communal affairs. This general control is strengthened by the authority granted to municipal secretaries or chief clerks who are state employees working in each of the communes. These "anomalous" state employees have since the 1920s assumed the position of inside watchdogs of communal governments (Cappelletti, 1963:262). In addition, each prefect controls the provincial fire brigade and is responsible for the maintenance of public order in his province.

Provincial officers of the state technical services have, since 1946, functioned as intermediaries between local and central offices and as monitoring agencies for the performance of local production units. Each province has, for example, a Purveyor of Studies, an Intendant of Finance, a provincial health officer, a Chief Engineer of the Civil Engineer Corps, and a Provincial Agrarian Inspector; each officer has in turn provincial inspectors assigned to make periodic visits to local production units to monitor performance of subordinate officials. In this capacity, each provincial state officer acts as a surrogate prefect. Much like "the other prefect," each field director renders account for the exercise of his authority only to his ministerial superiors. The delivery of public services that extends beyond the jurisdiction of a single "prefect" can thus be procured only through central directives aimed at coordinating provincial field services.

PERFORMANCE CRITERIA IN THE DESIGN AND EVALUATION OF INSTITUTIONAL ARRANGEMENTS

These local governmental institutions were not designed to be responsive to the preferences of diverse communities of people. To serve the need of localized or private interests was regarded as perverse or as "the first step to corruption." For private citizens to take a serious interest in public affairs or local problems was regarded as improper and even illegal. Only officials were presumed to concern themselves with public affairs or local problems (Pizzorno, 1971:90-91).

A uniform structure of governmental arrangements at the local level is presumed to serve the public interest of all citizens. The simplicity of the structure of communal government goes hand in hand with the clearly defined and concentrated authority and responsibility of communal officials. The business of checking on these officials is the exclusive prerogative of higher officials. These higher officials are in a position to monitor and evaluate the comparative performance of different communal governments and to determine the responsibility of communal officials if something goes wrong. The hierarchical organization of communal officials and professionally trained personnel is assumed to be the least costly or most efficient decision-making arrangement (1) for excluding localized or private interests in public decision-making, (2) for providing a uniform supply of public services, and (3) for monitoring the comparative performance of local governmental units. But what ultimately determines the single overall public interest is the central government. It is in this sense that institutional arrangements at the local level were designed to be responsive to the preferences or directives of central government authority.

After World War II, this particular set of institutional arrangements was still presumed to be a *neutral* device. The critical difference was a general recognition that the prevailing political concepts or the preferences of central government authority were "parliamentary" and not "Fascist" in nature. Thus, the artificers of postwar Italy assumed that the system of public administration which had been responsive to monarchical and Fascist leaders should now be responsive to republican leaders. The hierarchical arrangement that had been used to "unify" the country and to "shape" Italians in the image of imperial Rome would serve to advance democratic values and to bolster the republican regime. As a Minister of the Interior later observed in referring to the prefect as a symbol of that public administration:

> If the prefect did not exist it would be necessary to create him. His function is essential; he cannot be replaced. [Mario Scelba, 1960, quoted in Adams and Barile, 1972:120]

ELECTORAL LAWS AND PARTY ORGANIZATION

Before and after Fascism, access to the exercise of governmental prerogatives for the conduct of national and local government was dependent upon winning elections. The reiteration of the monocentric order after 1945 meant that those elected to exercise central governmental prerogatives would also be in a position to exercise a monopoly of authority over the supply of public services at the local level. The opportunities that derived from this structure of decision-

making arrangements provided the conditions for political entrepreneurs to organize and maintain national party organizations for
the purpose of capturing decision-making centers at the national and
local levels. Electoral competition was in turn enhanced and
constrained by electoral laws. Thus party organizations and electoral
laws provided the nexus between the structure of governmental
arrangements and the procurement and delivery of public services at
the local level.

The choice of electoral laws reflected the agreement between the
leaders of the major nongovernmental groupings that emerged during
the Second World War. These major groupings were the Demochristian party (DC), the Communist party (PCI), and the Socialist
party (PSI). These major parties were accompanied by small parties
that represented Social Democrats (PSDI), Republicans (PRI),
Liberals (PLI), Monarchists (PNM and PMP), and Neo-Fascists (MSI).
The strategic position acquired by these parties at the national level
together with the subordinate position of the structure of communal
government had the effect of insuring the dominance of these parties
at the local level and of precluding the development of independent
local organizations competing for municipal offices.

Before World War I, the electoral laws regulating elections to the
chamber of deputies and elections for communal offices in large
urban centers were based on a simple plurality vote in single-member
constituencies and city wards. Elections in towns with less than
5,000 inhabitants were instead based on a long party list with
plurality vote; a winning party list automatically received 80% of the
seats on the municipal council. Following the Second World War, the
electoral law for small towns remained unchanged; the electoral law
for towns with more than 10,000 residents was changed to an
at-large party-list system of proportional representation. Elections
for the chamber of deputies were based on a list system of
proportional representation in multimember districts or electoral
colleges.

Party-list systems with proportional representation implies that
representatives for a city in communal elections and for the chamber
of deputies in national elections are elected at large. The appeal to
the voters is to the voters of the city or the nation as a whole.
Elections become a popularity contest among the political parties.
Candidates are successful by virtue of their being on a party list and
among the winning proportion assigned to their political party. They
are intended to represent the overall public interest, not the discrete
interests of diverse constituencies or districts.

Proportional representation among party lists was in turn modified by a system of voter preferences within each party list. Voters would first choose which party they desired and then would be at liberty to choose not more than five among the names on that list as individual preferences. The number of seats assigned to each party would depend on the proportion of total votes received, but what determined the election of specific candidates from each party list in turn depended on the number of preference votes received.

THE OPERATION OF URBAN SERVICE SYSTEMS

According to the theory underlying the design of the Italian state, the consolidated unit of communal government and the local instrumentalities of the national bureaucracy were expected to yield a uniform level of public services among all neighborhoods within any particular city or town. The uniformity of local institutional arrangements was in turn expected to yield a uniform level of public services among all cities and towns. The monitoring of performance levels by administrative superiors was expected to insure that an appropriate level of services was provided to all citizens within and among all communes.

The work of scholars in public choice and of earlier analysts who applied economic reasoning to the analysis of public sector problems suggests consequences that radically deviate from those anticipated by the authors of Italian unification. In *Democracy in America* and *The Old Regime and the French Revolution,* Tocqueville suggested that a centralized bureaucratic system with a government elected by democratic vote would give rise to "democratic despotism" (1945, book 4; 1955:167-168). In his essay "Away With the Prefect!," Einaudi extended Tocqueville's analysis to indicate that under a rule of democratic despotism, local officials "merely learn to obey, to intrigue, to recommend and to seek influence" (1954:54). In turn, the analysis sustained by Gordon Tullock in *The Politics of Bureaucracy* (1965) suggests that large-scale bureaucracies will be subject to serious problems of institutional weakness and failure. Goal displacement and risk avoidance will be their characteristic features. In sum, these works suggest that a very large bureaucracy will:

1) become increasingly indiscriminating in its response to diverse demands, 2) impose increasingly high social costs on those who are presumed to be the beneficiaries, 3) fail to proportion supply to demand, 4) allow public goods to erode by failing to take actions to prevent one use from impairing others, 5) become increasingly error prone and uncontrollable to the point where public actions deviate radically from rhetoric about public purposes and objectives, and 6) eventually lead to circumstances where remedial actions exacerbate rather than ameliorate problems. [Ostrom, 1974:64]

On the basis of this analysis, we would expect the structure of institutional arrangements at the local level in Italy to fail to supply all public services in relation to a uniform standard of performance within and among all the communes. We would expect the pathologies associated with large bureaucratic organizations to be *the* normal characteristics of Italian urban service systems. Moreover, we would expect administrative superiors responsible for insuring appropriate levels of performance in relation to uniform standards of performance to experience loss of information and control in their efforts to monitor the output of public services. We would further expect citizen alienation to go hand in hand with poor performance.

Available evidence and inferential reasoning allow us to make an initial estimate of how the urban service delivery system operates in Italy. We begin, first, with an examination of the impact of electoral laws and the structure of governmental arrangements on the procurement and delivery of public services. Next, evidence drawn from Bologna and Naples is used to examine the impact of differences in controlling parties on the conduct of communal governments.

PARTY ORGANIZATIONS, ELECTIONS, AND GOVERNMENT COALITIONS

Both national and local elections manifest similar competitive struggles among political parties. Similar issues and problems are the subject of local and national electoral campaigns. (See, e.g., Pryce, 1957:1.) As parties strive to distinguish themselves ideologically from one another, they appeal to "great issues" involving the overall public interest of the nation. As a Turin newspaper noted during the 1951 local elections in that city:

No one has talked of the humble problems . . . of roads, school buildings, health services, . . . everyone has preferred to concern themselves with the Atlantic Pact, Korea, peace and war, western and eastern civilization and even with religion and atheism. [*La Stampa,* May 13, 1951; quoted in Pryce, 1957:11]

Communal elections are not designed to allow for the articulation of different solutions to "humble problems." A recent study of party activities in central Italian communities indicates that, during the 1960s, communal electoral appeals had changed little except in name—from the Atlantic Pact and Korea to Cuba and Vietnam (Stern, 1974).

Voting serves to emphasize the disjunction existing between participation in the electoral process and articulation of preferences for public services. In 1946 voting became compulsory. Compulsory voting provides citizens with a certificate of good conduct or civic duty which is required for most employment and applications. Electoral laws under the party list system encourage citizens to choose parties on the basis of their ideological positions on "great issues" rather than express their preferences for the humbler services.

At the same time, introduction of preference voting with a party list provided some opportunity for citizens to express their preferences for particular candidates. But this undermined an essential rationale for long party lists. Candidates of the same list are in competition with each other. In most parties two contests occur in each national and urban election: an interparty contest and an intraparty contest. The intraparty contest creates opportunities for successful candidates to use their influence to secure favorable decisions for electors from government authorities and administrative personnel. But in the Italian setting, where communal policies do not control official action in the bureaucracy, such interventions become a perverse manifestation of responsiveness or "a first step to corruption."

If the opportunities inherent in the electoral laws serve to enhance intraparty competition, the same opportunities place serious limitations on the ability of a *single* party organization to secure a legislative majority and serve as the government. Forming and maintaining a winning legislative coalition requires a different set of strategies from that used to appeal to the electorate. The different ideological positions emphasized during elections provide little or no incentive to different party leaders to move toward mutually agreeable positions. Government coalitions are formed among leaders of different parties or party factions, but such coalitions are organized and maintained by unanimous consent of the groups involved. Intraparty factions sometimes serve to bridge interparty differences.

Throughout the postwar years, the exercise of central government authority rested with a legislative coalition headed by the DC as the majority party and composed, until the Opening to the Left in 1963, of alternating small center parties. The condition of unanimous consent for the establishment and maintenance of these national coalitions gave rise to holdout strategies generally associated with a rule of unanimous consent (Buchanan and Tullock, 1962; Sproule-Jones, 1974). During the 1950s, Demochristian legislative leaders also relied from time to time on the "external" and "uncovenanted" votes of Monarchists and Neo-Fascists in efforts to minimize their dependence on the small center parties.

The formation of central government coalitions based upon unanimous consent is reflected in efforts to replicate the same majority coalitions at the local level. Party and faction leaders need to control both sets of coalitions in order to retain their strategic positions. The loss of a coalition at one level undermines the coalition at the other level. (See, e.g., Allum, 1973:244; Barnes, 1974; Fried, 1973:147.)

Following the 1956 local elections, for example, national party and faction leaders composing the central coalition faced the task of negotiating the formation of communal coalitions in more than 200 cities (Pryce, 1957:106-108). During the postwar period, both right and left party organizations were successful in capturing communal governments in cities such as Naples and Bologna. These communal governments took up countervailing positions vis-à-vis the Demochristian party and the successive national coalitions. In this period, the relationship of each communal government with the central government ranged from one of mutual support to mutual antagonism.

Administrative officials responsible for monitoring performance came to use the position of each communal government vis-à-vis the central government as a critical performance variable. A condition of mutual support between the central government and the communal government is used to facilitate informal arrangements. A condition of mutual antagonism is used instead to scrutinize the conduct of communal government. The degree of central control over local authorities is so substantial that, if, for example, prefects who "interpreted the law punctiliously and conscientiously intervened in the activities of local authorities in every case in which (they) are entitled to do so, local authorities could hardly operate at all" (Cappelletti, 1963:261-262). How this complex interweave of party organizations, elections, and government coalitions affects the operation of communal governments will be examined in the case of Bologna and Naples.

THE CONDUCT OF COMMUNAL GOVERNMENTS

Communal government authorities have no essential control over the delivery of public services supplied by the national bureaucracy. Given the uniform structure of communal governments, what difference does the control of communal governments by different party organizations make in the conduct of communal governments? Available evidence on tax policies and in the conduct of government in Bologna and Naples provides a tentative answer to this question.

Evidence about communal tax policies comes from a Carlo Cattaneo Institute study. Investigators used the communal sales tax and the family tax as indicators of the effect of party control on communal revenue policies. Sales or consumption taxes fall largely on items of necessity and are considered instruments of conservative fiscal policy. The family income tax is generally considered an instrument of progressive fiscal policy. Elected communal officials have discretion in the levy of these sources of revenue. Investigators expected that, given left ideology, Communist communal governments would show relatively high per-capital family income tax levels and relatively low sales tax levels. The reverse was expected in the case of Demochristian and right communal governments (Galli and Prandi, 1970:241-243).

The research found, however, that "it was not possible to distinguish Communist from (Demochristian) taxation policies" (Galli and Prandi, 1970:242). Both Communist and Demochristian communes showed a heavy reliance on family income tax. Communist-controlled Bologna, for example, had a higher percentage of revenues from the sales tax (35%) than the national average (33%) or the cities of Milan (30.1%) and Rome (30.4%), which were controlled by the Demochristians. This finding led the analysts to note that the "Communist party's program for increasing the family tax and reducing the sales tax had apparently not applied in Communist-administered cities" (Galli and Prandi, 1970:243). The general conclusion of this research was that "differences in budgetary policies were not associated in any systematic or consistent way with the control of communal governments by different party organizations" (Galli and Prandi, 1970:241; but see Fried, 1971).

Similar tax policies may, however, mask quite different patterns of expenditures and quite different levels of performance. Evidence drawn from Robert H. Evans' *Coexistence: Communism and Its Practice in Bologna, 1945-1965* (1967) and P.A. Allum's *Politics and Society in Post-War Naples* (1973) provides an opportunity to compare the conduct of communal affairs in Bologna and Naples. Since the end of the Second World War, Bologna has been administered by successive Communist-Socialist coalitions led by Communist mayors. During the 1950s Naples was administered by a Monarchist-Neo-Fascist coalition led by a Monarchist mayor, and during the 1960s it was administered by successive center-left coalitions led by Demochristian mayors.

Empirical observations in both cities reveal some common operational characteristics. The structure of communal institutions that were expected to yield a uniform level of services is used instead as an instrument of party organizations. Irrespective of party labels, elected officials of both communal governments take advantage of the opportunities inherent in the structure of communal government to enhance the objectives of their party organizations. Citizens' access to municipal services occurs more as a party or personal favor than as a right to equal municipal services.

The use of communal institutions as instruments of party organizations takes several forms in both cities. Both Communist and Monarchist mayors place their respective party members in communal departments and enterprises (Evans, 1967:164; Allum, 1973:163-164) and provide their supporters with "letters of recommendation" for easy access to communal services (Evans, 1967:74, 204-206, 211-212; Allum, 1973:176-177). They also use their authoritative positions to grant private developers "exceptions" to zoning regulations (Fried, 1973:254; Montanelli et al., 1965:391; Allum, 1973:286) and to award municipal contracts ranging from food catering services to public works projects to their respective party members and supporters (Evans, 1967:132, 134, 163; Allum, 1973:285; see also Passigli, 1963:729-730). Thus different party cards are similarly "bread cards."

The mutually antagonistic positions of the Demochristian and Communist parties are reflected in "rigid" prefectoral surveillance over Bolognese communal affairs (Evans, 1967:163). Even though rigid control is exercised, it is difficult for the prefect or prefectoral officials to exert effective control over the conduct of *all* Bolognese communal affairs. This problem is especially evident in the case of municipal contracts awarded to Communist production cooperatives (Evans, 1967:314). In the building of an underground passage, for example, a contract clause required no traffic interruption and invoked a one million lire penalty for every day of interruption. This contract clause was waived only after that public work had been awarded to a Communist cooperative (Evans, 1967:163, note 31). In turn, cooperative leaders and individual members are required to turn over part of their earnings to party officials (Evans, 1967:110-111, 163; see also Passigli, 1963:729). By the late 1950s, control of the instrumentalities of communal government in Bologna and neighboring communities had transformed the Communist party into the largest private enterprise in the province (Evans, 1967:134; see also Galli and Prandi, 1970:162).

Between 1952 and 1953, the relationship between the Monarchist communal government of Naples and the central government was characterized by partial antagonism. The 1953 national elections altered that relationship. In those elections the Demochristian party suffered a sharp loss of legislative seats; the Monarchist party instead trebled its legislative representation from 14 to 40 deputies. In an effort to sustain the legislative viability of the ruling coalition, Demochristian leaders came to rely upon the mayor of Naples, Achille Lauro, for securing the "uncovenanted votes" of Monarchist deputies. Lauro was also a deputy and a national leader of the Monarchist party. In 1954, Demochristian provincial leaders also sought and secured Lauro's help in order to wrest from left control several communes in the Neapolitan province (Allum, 1973:285-286; Fried, 1963:258).

Thus the legislative requirements for sustaining the central government coalition and for wresting communal governments from left control placed the mayor of Naples in a position to pursue the opportunities inherent in the structure of communal government under a condition of immunity from prefectoral control. In addition, the same conditions gave the mayor of Naples easy access to special funds for public works in Naples and to almost unlimited communal spending powers (Allum, 1973:286).

Monarchist officials at the top of communal departments and enterprises found it relatively easy to procure funds for capital expenditures on transportation facilities and water supply systems. However, they experienced serious difficulties in checking on how their subordinates discharged public services in person-to-person relationships on behalf of the Monarchist party. In an effort to reduce the frustration and cynicism of party members and supporters with bureaucratic unresponsiveness, the mayor of Naples began to intervene personally in the conduct of communal departments and enterprises. These "personal interventions" served to make Lauro an extremely popular mayor. They also served to compound the problems already associated with the existing bureaucratic practices of city-wide public organizations. The personal interventions as well as "general irregularities" were brought to the attention of judicial and prefectoral authorities by members of the communal opposition. But the need to insure legislative support of the central government coalition sustained the Monarchist communal government.

By 1958, the approaching national elections served to alter this particular relationship of the central government with the Naples communal government. Efforts of Demochristian leaders now shifted from legislative to electoral requirements. A few months before the 1958 national elections, the Minister of the Interior dissolved the Naples communal government on grounds of "grave irregularities." While a prefect took charge of communal affairs, the Monarchist mayor was dismissed and faced criminal charges.

It was not until about three years later that communal affairs were turned over to elected officials. Now Demochristian leaders replaced Monarchist leaders in control of the Neapolitan commune. The conduct of communal affairs during the 1960s once again became subject to the requirements of national politics. The previous communal "irregularities" were now magnified by the full correspondence between leaders of the central government and leaders of the communal government (Allum, 1973:307-324). Under these conditions judicial proceedings against Achille Lauro were postponed indefinitely.

These events sharpened the general dissatisfaction already existing among ordinary Neapolitans about the instrumentalities of government and the provision of public services. Several surveys of citizens' perception of communal government conducted in the 1960s reveal that the high level of citizen alienation was grounded in the knowledge that ordinary Neapolitans have "no more control over the caprices of human authority than [they do] over the uncertainties of the physical universe" (Allum, 1973:98). But, as we have seen, public officials also experience serious difficulties in their efforts to exercise control over communal institutions.

By contrast, the interweave between national and local Communist party politics joined with the competitive relationship between the central and communal governments to subject the conduct of Bolognese communal affairs to quite different requirements of national party politics. For example, between 1946 and 1956 the Communist party program included a revolutionary takeover of the country. However, the opportunity to transform the control of communal governments into positions of opposition to the Demochristian-dominated central government led national Communist officials to call for local self-government and the abolition of prefects. An eventual takeover of the country was expected to start from the communes.

In Bologna the application of this strategy took several forms. Communal officials identified the PCI with local self-government and emphasized how the structure of communal government "suffocated" rather than facilitated municipal initiative on behalf of common interests shared by urban residents (e.g., Giuseppe Dozza, 1947 and 1951; quoted in Evans, 1967:151). In order to retain some measure of autonomy from the central government, Bolognese communal officials relied very little on state subsidies in meeting communal expenditures and in balancing the budget (Evans, 1967:37-38, 154). This may account for the relatively high communal sales tax cited earlier in the discussion of the Carlo Cattaneo Institute study of communal tax policies. However, "Bologna's status as one of the few Italian towns with a balanced budget provided excellent propaganda" for the Communist party (Evans, 1967:38). It was in this sense that by the early 1950s the Communist press began to portray Bologna as "Italy's model city."

After 1956, in an attempt to minimize exposure to the vicissitudes of Eastern European communism, PCI leaders officially abandoned the call for a revolutionary takeover of the country in favor of an "Italian Way to Socialism." This new party program accepted the existing institutional arrangements. The switch in national party strategy led Bolognese communal officials who had earlier stressed the "suffocating" nature of the structure of communal government to recognize the same structure now as "the organic expression of the State" (e.g., Giuseppe Dozza, 1965; quoted in Evans, 1967:151). There were no more calls for the abolition of the prefectoral system.

The Italian Way to Socialism led to several changes in fiscal policy and the organization of Bologna communal government. The policy of a balanced budget, for example, gave way to a policy of state subsidies and deficit spending. Between 1958 and 1964 several capital projects were constructed and some 290,000 square meters of land were purchased for the construction of low-cost housing and resold at a low price to Bolognese residents. The new public works projects provided capital intensive services for local residents and new employment opportunities for members of Communist cooperatives. However, they also brought Bologna to the same level of high deficit spending as Naples and most other Italian towns. Deficit spending was, in turn, accompanied by new central government constraints in the form of surveillance by officials of the Central Finance Commission responsible for deficitary communes (Evans, 1967:38, 160-162).

In 1964, under a 1915 dormant national law, the city was divided into 15 "zones," each with a "little city hall" and a district council. Each little city hall was expected to serve as an administrative service center. Each district council, composed of 20 members chosen by the municipal council according to the distribution of seats in the communal council and presided over by a mayor's representative, was expected to serve as a consultative organ between local residents and communal officials (Evans, 1967:158-159).

The establishment of district councils reinforced the image of Bologna as "Italy's model city" by anticipating the establishment of similar councils in other cities. However, neighborhood councils had been originally proposed by the Demochristian mayoral candidate in the 1956 Bolognese election. By 1965, when Evans' study ends, the district councils were minimally operational. Since then, much of the prevailing knowledge about their operation is based upon interviews with the mayor of Bologna. (See, e.g., Dragone, 1975:169-178; L'Espresso, Rome, April 20, 1975, p. 99; but see Centro Studi Economici, 1972.) Thus it is not clear whether these district councils have diminished the relative isolation of Bolognese residents from communal government (Evans, 1967:185). These changes have, however, served as a model for recent attempts on the part of other Italian communal officials to improve service delivery. These and other reforms aimed at decentralizing governmental power in Italy are described below.

RECENT INSTITUTIONAL CHANGES

During the course of the 1950s, calls for urban reforms became part of a more general call for a reform of the state. These calls for reform raised the problem of how parliament—the only governmental unit with the authority to make structural changes at all levels—could initiate basic reforms when the parliamentary groups that form governing coalitions have incentives to maintain the dominance of central authorities. Thus there could be no change contrary to the principles of centralized control over communal public services. Only conditions that threaten the survival of the parliamentary regime itself would force parliamentary groups to accede to basic reforms in the structure of government.

The Opening to the Left in 1963 was expected to initiate a policy of structural reforms. Instead, it extended the privileged position of the Demochristian party to the Socialist party (PSI). It was not until the "hot autumn" of 1969, which threatened the very existence of the parliamentary regime, that the necessary consensus for structural changes among leaders of the ruling coalition was created. This consensus resulted in 1972 in the establishment of 15 regional governments in addition to those already at work on the islands and in the Alpine border areas.

Well before 1972, however, communal officials of large cities, faced with central government inattention to urban reforms, began to search for ways to meet local demands for structural changes. Communal decentralization in Bologna provided an example of what communal officials could do within the existing legal framework. By the early 1970s, 100 cities, including at least 66 of the 93 provincial capitals, had adopted, with some variations, the Bolognese type of communal decentralization.

Both types of institutional change were expected to improve the performance of urban service systems and to bring government closer to the citizens. Yet the events described in the beginning of this essay suggest that communal and central government decentralization has not produced the desired results. Why?

Unfortunately there is a dearth of "before-after" evaluative studies of the impact of communal and central government decentralization. However, one way to proceed in an initial, if tentative, evaluation is to describe these structural changes. A description of recent institutional changes can serve to reveal how these changes were expected to correct earlier sources of institutional weakness and failure. We can then draw upon the limited evidence available to discuss some of the short-term effects that these changes have had.

COMMUNAL DECENTRALIZATION

The enabling legislation for communal decentralization is based upon the communal and provincial law of 1915. By 1910, the failure of large communal governments to produce expected results had become a national concern. This concern was expressed in a general recognition that (a) within large communes there were subcommunities with diverse interests and problems, (b) the structure of communal government was not sensitive to these problems, and (c) these problems might well be solved more effectively by local residents through some kind of district government such as that which had existed in Naples before 1860. National legislators, however, had difficulty reconciling these principles of self-government with the principles of control by central authorities. The solution was to recognize the existence of "smaller areas of common interests" or city districts but to give them a "voice" through mayor's representatives and district "associates" *(aggiunti)*—or in essence, to create a "prefectoral" system at the communal level (Crea, 1965).

The vicissitudes connected with the First and Second World Wars precluded the implementation of this legislation. The first application took place in Bologna in 1964. During the latter part of the 1960s, other communal governments followed the Bolognese example. The essential characteristics of what became known as the "Bologna model of decentralization" were as follows:

(1) Each city was divided in zones or districts primarily on the basis of their historic district boundaries dating to pre-1860 times. Each district had a set of "associates" or council, a "little city hall," and a mayor's representative.

(2) Each district council was composed of 20 local residents chosen by the municipal council according to the proportion of party seats on the municipal council.

(3) Each district council had only the power to make recommendations or suggestions to the municipal council on local problems. Recommendations or suggestions were not formally binding on city officials. District councils were not even permitted to hire their secretary or clerk.

(4) Each "little city hall" served as a decentralized service center for the municipal bureaucracy. The administrative director of each service center was also the secretary or clerk of the district council.

(5) A mayor's representative, not necessarily from the same district, directed and presided over each district council. He also has responsibility for coordinating and monitoring the activities of the decentralized municipal departments.

Thus district councils reproduce in microcosm the composition of municipal councils. Mayors' representatives have become known as "mayors' prefects" (Centro Studi Economici, 1972:12). Administrative decentralization reiterates at each district level the hierarchic structure and operation of communal departments. There is no basic change in communal decision-making authority. (But see Stoppino, 1975:489-494.)

Among the very large urban cities that adopted the "Bologna model," with some variations, were Rome in 1966, Milan and Naples in 1968, and Genoa in 1969. In most cases, the districts created were themselves relatively large, because the use of historic boundaries as a criterion for districts failed to take into account increases in population. For example, the average number of residents for each of the 12 Roman districts is 210,000; the average for each of the 20 Milanese districts is 85,000. In Naples three new districts were added to the 12 pre-1860 districts, but the average number of residents for each area remains over 80,000. The operation of communal decentralization thus involves relatively large-scale organizations (Dente, 1974:184-185).

Critics from extraparliamentary groups charge that district councils simply serve to strenthen the strategic position of party organizations and to undermine and preempt neighborhood efforts by local residents (Della Pergola, 1974). Local elected officials reject these charges and contend instead that this type of communal decentralization permits a form of citizen participation that allows public officials to respond more effectively to local problems (Dragone, 1975:13-28).

A study of the operation of four Milanese district councils, three of which had less than 75,000 residents, suggests that district councils have served neither to strengthen the strategic positions of party organizations nor to undermine or preempt neighborhood efforts such as tenant associations and youth groups (Daolio, 1971). The investigator also found that district councils were either not consulted on matters of district concern or consulted but without having their suggestions incorporated in communal policies. As a result, the four district councils presented "alternative proposals for the organization and operation of district councils" to the Milanese communal government (Daolio, 1971:30-31).

These alternative proposals included direct elections of district council members, power to hire district clerks and other officials, independent decision-making authority in matters of district or neighborhood concern, and "participation" with the communal government in levying and controlling communal taxes of district residents. These demands, however, went beyond the legal framework of the 1915 law and, as a result, they were never adopted. The general conclusion of the study was that the establishment of district councils revealed both the existence of a large number of neighborhood residents willing to act on behalf of common interests and the lack of governmental structures to facilitate such collective efforts (Daolio, 1971:31; see also Dente, 1974:194-196; Stoppino, 1975:504-505).

CENTRAL GOVERNMENT DECENTRALIZATION

The second major innovation in governmental organization undertaken in the recent past has been the creation of regional governments. Regional government was initially authorized in an effort to accommodate Sicilian claims to an inherent right of self-government that resurfaced during the Second World War. By 1944, central government officials had recognized the failure of efforts to create national unity through administrative measures and the existence of special regional problems in the provision of many public services. The same officials had, however, serious difficulties in meeting Sicilian claims to self-government without changing the unitary form of the state. The solution was a special provision for regional government in Sicily—leaving the basic structures of the unitary state unchanged.

The exceptional nature of this kind of central government decentralization was implied in the designation of the structure of Sicilian regional government as an "extraordinary" institutional arrangement. This special arrangement was subsequently used to meet regional claims of independence or foreign threats of annexation in Sardinia, Francophone Val d'Aosta, and Germanophone Alto Adige or South Tyrol. In 1963, following the return of the city of Trieste to Italian rule, a fifth extraordinary region was created near the Yugoslav border.

The organization of the five extraordinary regions was patterned upon a unicameral system of parliamentary government with proportional representation, long party lists, and preference voting. With some variations such as the use of French and German, all five regional governments had the following powers:

(1) To exercise "limited exclusive" jurisdiction in such areas as land reform, public works, and urban planning and "complementary" or concurrent jurisdiction with the central government in such areas as public health and hospital care, commerce, and water resources.

(2) To adapt the details of national laws to the specific needs and conditions of each region whenever such "integrative" powers were delegated by the national parliament.

(3) To initiate proposals for submission to parliament concerning matters of regional interests outside their own legislative jurisdiction.

(4) To levy and collect regional taxes limited essentially to the use of public soil or property, to receive a share of certain national taxes collected in each region, and to borrow money for specific public projects. About two-thirds of regional revenue, however, came from annual grants from the central government.

(5) To establish regional ministries or technical services.

(6) To monitor the performance of communal governments through the establishment of provincial control commissions responsible to "regional ministers of interior."

Regional autonomy did not, however, imply a diminution in the sovereignty of the national parliament. National legislation automatically took precedence over regional legislation. For example, regional "limited exclusive" jurisdiction could take place only when national legislation contained provisions limiting its sphere of application. General responsibility for monitoring the legality and merit of acts of regional governments rested with state commissioners or super-prefects posted in each regional capital. The central government retained authority over (1) the substance of regional statutes, (2) the decision-making powers assigned to regional governments, and (3) monitoring of the performance of regional governments (Woodcock, 1967, 1969).

Thus the structure of authority relationships inherent in the conduct of national affairs was reflected in the conduct of regional affairs. At the same time, the five regional authorities had no essential control over the national bureaucracy. Regional services would have to be carried out in relation to formal rules and regulations promulgated by the national government. In turn, communal authorities had no essential control over either the national or regional bureaucracies. Communal services now would have to be carried out in relation to formal rules and regulations promulgated by both the central and regional governments.

The establishment of 15 "ordinary" regions in 1972 follows, with some variations, the essential features of the five extraordinary regions. Variations between the two sets of regional governments involves primarily the assignment of different legislative powers. Ordinary regional governments are not, for example, empowered to exercise "limited exclusive" jurisdiction over some policy areas or to submit legislative proposals to parliament. As we have seen, however, "limited exclusive" jurisdiction by extraordinary regions can take place only when national legislation so specifies. At the same time, the new regional governments have no control over the national police system or the provision of public security. Local *polizia* officers continue to be subject to a command structure in the Ministry of the Interior; prefects continue to exercise general responsibilities for peace and security in each province. Hence the extension of regional government to the rest of the Italian peninsula has served in effect to underscore the subordinate position of regional authorities vis-à-vis the central government and the national bureaucracy. (See, Barile, 1975:401-411, 415.)

Leaders and legislators of the major political parties were confident that this structure of regional government would correct the malfunctioning of the national bureaucracy and bring government closer to the citizens. A student of Italian urban service systems has attempted to elucidate the basis of this confidence by examining the positions taken by the various political parties on the implementation of regional government (Villani, 1972). He has found that the national debate on structural changes was "burdened by ideological and righteous *(politici)* elements that in many ways transcended the question of organizational efficiency designed to provide the maximum of collective goods to given resources and to match as closely as possible the preferences of consumers of those goods" (Villani, 1972:3). The confidence expressed by Communist, Demochristian, and Socialist officials rested upon an inadequate consideration of the potential effects of the structural change being introduced. Consequently, no one saw the need to examine the impact of regional governments in areas where they had been operational for some 20 years. Available evidence about the operation of regional government in Sicily and in some of the new regions provides but a glimpse of some of the consequences that institutional changes burdened by ideological or righteous elements may have.

As we have seen, access to the exercise of regional government authority is dependent upon winning elections. With the first regional election held in 1947, the Demochristian party emerged as the principal ruling party in Sicily. As a result, the formation and conduct of successive Sicilian regional governments became subject to electoral and legislative requirements of national government coalitions (Pantaleone, 1972). In turn, the formation and conduct of communal government in such cities as Catania and Palermo became subject to electoral and legislative requirements of both national and regional government coalitions (Capurso, 1964).

At the same time, the establishment of regional ministries to coincide with central government ministries was accompanied by a transfer of state administrative personnel to the regional bureaucracy. Most of these high administrative officials came from disbanding the ministry for colonial affairs. By the 1950s, intergovernmental relations between regional ministries and state technical services evidenced the same problems that afflicted intergovernmental relations between communal departments and state technical services. This bureaucratic dominance was in turn compounded by resentment among Sicilian administrative personnel over the high bureaucratic positions occupied by former colonial administrators (Serio, 1966).

Faced with the requirements of national party politics and the problems of bureaucratic control, Demochristian regional officials proceeded by 1953 to transform the regional bureaucracy into a work place for party functionaries. The informal and personal networks that developed cut across ideological and intergovernmental barriers and became fairly responsive to the articulation of demands for services by citizens and communal officials. Thus patterns of intergovernmental relations between central, regional, and communal authorities took on the appearance of a polycentric rather than a monocentric ordering. This polycentricity accrued, however, from a logic of corruption and *not* from a design of self-governing, independent levels of government with overlapping jurisdictions (Ostrom, 1972:7).

In 1958, some regional Demochristian deputies refused to recognize the party central executive's claim to designate the head of the Demochristian regional government and were expelled from the party. These events led to a constituency revolt against party discipline which in turn gave rise to an independent Sicilian Catholic Social party. Leaders of this party proceeded to form a coalition government with Communists, Socialists, Monarchists, and Neo-Fascists. Whereas on the Italian peninsula, this "unnatural alliance" was seen as proof of Sicilian political immaturity, in Sicily it was seen as an expression of a common concern for Sicilian home rule. This identity of interests gave rise to the possibility of a political stalemate between the central government and the regional government that would have provided the conditions for a reformulation of the structure of Sicilian autonomy. By 1960 the "unnatural alliance" collapsed. In 1961 the legislative support of the Socialist regional party organization helped to reestablish control by the Demochristian party in the formation and conduct of Sicilian regional government.

By the middle of the 1960s the organizational chart of the regional bureaucracy resembled an inverted pyramid. Some 400 technical personnel at the middle and bottom level of this inverted pyramid expressed strong dissatisfaction about regional arrangements. Organized as a club (Associazione Culturale Funzionari Regionali), these regional bureaucrats supplemented work protests with reports about administrative practices. They proposed a system of democratic public administration based upon a restructuring of the regional bureaucracy in terms of "operative units" or "service bureaus" and a restructuring of incentives for administrative personnel that would follow from the less hierarchical and more service-oriented "operative units." The success of this alternative system of regional administration would depend in turn on the restructuring of the national public administration being proposed by the national minister for bureaucratic reforms. But the resistance to changes among other members of the regional bureaucracy and the absence of changes in the national bureaucracy undermined these efforts.

Thus a constituency revolt within one party organization was insufficient to establish regional autonomy in the formation of regional governing coalitions. Work protest among some regional bureaucrats was insufficient to change the operating characteristics of the regional bureaucracy. As long as the hegemony of central party headquarters prevailed over a majority of regional party organizations and as long as the central bureaucracy prevailed over the regional bureaucracy, it was difficult to alter the structure of regional government. By 1970, when regional government was being extended to the rest of Italy, the discrepancy between the public performance of Sicilian regional government and the public rhetoric about regional purposes and goals was difficult to reconcile.

A study by Norman Kogan (1975) provides an assessment of the impact of the new regional governments on the structure of power within the parties. The research was carried out in 1973, a year after the 15 ordinary regions became operational. By that time, the majority of regions were governed by Demochristian-led center-left coalitions, while three regions in the so-called Red Belt area were governed by Communist-led left coalitions. Kogan found that, in spite of some dissatisfaction among regional party officials about central party directives, the national party or faction leaders retained essential authority over regional party organizations. The control of regional government by some Demochristian party organizations did not radically alter the authority structure within that party that existed prior to 1970. In the case of the Communist party, several PCI regional secretaries and the head of a Communist-led regional government were named to that party's central committee, but "democratic centralism" remained to assure central control over party policies. The general conclusion of the study was that "resistance to sharing power inside the parties is strong" (Kogan, 1975:403).

In the course of his analysis, Kogan also noted the tendency among central government leaders and bureaucrats "to treat regions as field offices of the national administration, executing orders and policies handed down from Rome" (p. 403). This impression is supported by a detailed study of the national ministries most affected by the extension of regional government (Amato, 1974). The transfer of jurisdiction from the national ministries of interior, public works, health, transportation and tourism to their regional counterparts has been accompanied by rules and regulations to insure the "imperial" position of the central bureaucracy (Amato, 1974:482, 485, 489). The tendency "to treat regions as field offices of the national administration" or to insure the "imperial" position of the central bureaucracy conforms, however, to the design criteria of regional government. This state of affairs has led the president of the Tuscan region, Lelio Lagorio, to acknowledge recently that "a truly regional government still remains to be constructed" (quoted in *Il Mondo,* Milan, May 22, 1975, p. 67).

Thus much like communal decentralization, the hopes expressed about central government decentralization have failed to meet the test of experience. If this initial assessment of recent institutional changes is valid, then the question of "why the persistent failure in the performance of urban service systems?" raised at the beginning of this essay can be answered in terms of a highly centralized structure of governmental arrangements.

PROGNOSIS FOR THE FUTURE

In the course of the second half of the 19th century, Sicilian publicists such as Napoleone Colajanni and Edoardo Pantano likened proposals to decentralize central and communal government authority to attempts to shorten the handle of a hammer when the hammer of central bureaucracy itself was the problem (Ganci, 1973:51). These analysts were skeptical that such efforts would correct the malfunctioning of the national system of public administration. At the same time, they were aware that the transformation of a highly centralized state into a state based upon a different design could occur only over a long period of time. Hence, they came to view proposals to shorten the handle of the bureaucratic hammer as opportunities that might provide the conditions which would lead to a transformation of the Italian state into a different system of government. This different system of government, they anticipated, would be based upon an extension of self-government to neighborhoods, communes, regions, and the nation. These Sicilian publicists used the writings on democracy in America by Madison and Hamilton and Tocqueville to support their contention that such a design of government was conceptually and operationally feasible. (See Colajanni, 1883 and 1896, cited in Ganci, 1973:51-69, 261-285.)

If the scourge of war can now be avoided for several generations, the establishment of district councils and regional government may provide the conditions for a gradual transformation of the Italian public service system to emphasize *service* rather than *control.* Proposals for the conduct of district councils advanced by the four Milanese district councils and the Sicilian efforts to alter the terms and conditions of Sicilian regional autonomy suggest that the political knowledge necessary to affect changes in the structure of organizational arrangements does exist. If other district councils in other cities succeed in similar efforts and if a revolt of constituencies succeeds in restructuring the political parties, opportunities may then exist to transform the imperial state into one that is more service-oriented.

As long as the hegemony of central party headquarters prevails over local and regional party organizations, however, it is difficult to anticipate changes in the central government dominance over the service structure in communal and regional governments. Competitive rivalry between the Demochristian party and the Communist party need not, as the case of Bologna suggests, radically alter these conditions. The leaders of each party have incentives to preserve the prerogatives of the central government as soon as they become the government and can exercise those prerogatives. They have no incentives to alter the structure of government to facilitate self-government in the different neighborhoods, communes, and regions of Italy.

REFERENCES

ADAMS, J.C., and BARILE, P. (1972). The government of republican Italy. Boston: Houghton Mifflin.
ALLUM, P.A. (1973). Politics and society in post-war Naples. Cambridge, Eng.: Cambridge University Press.
AMATO, G. (1974). "Gli apparati centrali e le regioni." Pp. 478-491 in S. Cassese (ed.), L'amministrazione publica in Italia. Bologna: Il Mulino.
BARILE, P. (1975). Istituzioni di diritto pubblico. Padova: CEDAM.
BARNES, S.H. (1974). "Decision-making in Italian politics." Administration and Society, 6(August):179-204.
BUCHANAN, J.M., and TULLOCK, G. (1962). The calculus of consent. Ann Arbor: University of Michigan Press.
CAPPELLETTI, L. (1963). "Local government in Italy." Public Administration, 41:247-264.
CAPURSO, G.L. (1964). "Cronache amministrative: Palermo." Nord e Sud, Naples, July, pp. 52-56.
CATTANEO, C. (1945). Stati uniti d'Italia (N. Bobbio, ed.). Torino: Chiantore.
CAVOUR, C. (1949). Carteggi: La liberazione del mezzogiorno e la formazione del regno d'Italia (4 vols.). Bologna: Zanichelli.
Centro Studi Economici (1972). La sinistra al governo, Bologna: Riflessione e proposte alternative sulla gestione del potere in una città democratica. Bologna: Edizioni del Centro Studi Economici.
CREA, P. (1965). "Sul decentramento amministrativo di grandi centri urbani." Pp. 62-73 in G. Maranini (ed.), La regione e il governo locale (vol. 2). Milan: Edizioni di Comunità.
DAOLIO, A. (1971). "L'esperienza Milanese dei consigli di zona." Economia Pubblica, Milan, October, pp. 23-32.

DELLA PERGOLA, G. (1974). Diritto alla città e lotte urbane. Milano: Feltrinelli.

DENTE, B. (1974). "L'organizzazione di governo degli enti locali: Rassegna di proposte ed esperienze 1961-1973." Rivista Trimestrale di Diritto Pubblico, Milan, 1:150-199.

DRAGONE, U. (ed., 1975). Decentramento urbano e democrazia: Milano, Bologna, Roma, Torino, Pavia. Milano: Feltrinelli.

EINAUDI, L. (1954). "Via al prefetto!" Pp. 52-59 in L. Einaudi, Il buongoverno: Saggi di economia e politica, 1897-1954 (E. Rossi, ed.). Bari: Editori Laterza.

EVANS, R.H. (1967). Coexistence: Communism and its practice in Bologna, 1945-1965. Notre Dame, Ind.: University of Notre Dame Press.

FERRARA, F. (1949). "F. Ferrara a Cavour: Brevi Note sulla Sicilia (July)." Pp. 296-305 in C. Cavour, Carteggi: La liberazione del mezzogiorno e la formazione del regno d'Italia (vol. 1). Bologna: Zamichelli.

――― (1965). "Unione non unità." Pp. 325-331 in F. Ferrara, Opere Complete (vol. 1, part 1; F. Sirugo, ed.). Roma: Associazione Bancaria Italiana e della Banca d'Italia.

FRIED, R.C. (1963). The Italian prefects: A study in administrative politics. New Haven, Conn.: Yale University Press.

――― (1971). "Communism, urban budgets and the two Italies: A case study in comparative urban government." Journal of Politics, 33(November):1008-1051.

――― (1973). Planning the Eternal City: Roman politics and planning since World War II. New Haven, Conn.: Yale University Press.

GALLI, G., and PRANDI, A. (1970). Pattern of political participation in Italy. New Haven, Conn.: Yale University Press.

GANCI, S.M. (1973). Da Crispi a Rudinì: La polemica regionalista, 1894-1896. Palermo: S.F. Flaccovio.

ISAP (1962). Archivio dell'Istituto per la Scienza dell'Amministrazione Pubblica (vol. 1). Milano: Giuffrè.

KOGAN, N. (1975). "Impact of the new Italian regional government on the structure of power within the parties." Comparative Politics, 8(April):383-406.

LEVI, C. (1963). Christ stopped at Eboli (F. Frenaye, trans.). New York: Noonday Press.

MONTANELLI, I., CAVALLARI, A., OTTONE, P., PIAZZESI, G., and RUSSO, G. (1965). Italia sotto inchiesta: "Corriere della sera" 1963/65. Firenze: Sansoni.

OSTROM, V. (1972). "Polycentricity." Paper delivered at the annual meeting of the American Political Science Association, Washington, D.C., September 5-9.

――― (1974). The intellectual crisis in American public administration. University: Alabama University Press.

PANTALEONE, M. (1972). L'industria del potere. Bologna: Cappelli.

PASSIGLI, S. (1963). "Italy." Journal of Politics, 25(November):718-736.

PETRACCHI, A. (1962). Le origini dell'ordinamento comunale e provinciale Italiano: Storia della legislazione Piedmontese sugli enti locali dalla fine dell'antico regime al chiudersi dell'età Cavouriana, 1770-1861 (2 vols.). Venezia: Neri Pozza.

PIZZORNO, A. (1971). "Amoral familism and historical marginality." Pp. 87-98 in M. Dogan and R. Rose (eds.), European politics: A reader. Boston: Little, Brown.

PRYCE, R. (1957). The Italian local elections, 1956 (St. Anthony's Papers, no. 3). London: Chatto and Windus.

ROTELLI, E. (1973). "Le transformazioni dell'ordinamento comunale e provinciale durante il regime fascista." Pp. 73-156 in S. Fontana (ed.), Il Fascismo e le autonomie locali. Bologna: Società Editrice Il Mulino.

SERIO, E. (1966). "Burocrazia in Sicilia." Nord e Sud, Naples, June, pp. 51-57.

SPROULE-JONES, M. (1974). "An analysis of Canadian federalism." Publius, 4(fall):109-136.

STEINER, H.A. (1939). "Italy: Communal and provincial government." Pp. 305-380 in W. Anderson (ed.), Local government in Europe. New York: D. Appleton Century.

STERN, A.J. (1974). "The Italian CP at the grass roots." Problems of Communism, 23(March-April):42-54.

STOPPINO, M. (1975). "Decentramento comunale e partecipazione politica." Amministrare, Milan, no. 4, pp. 489-505.

TOCQUEVILLE, A. de (1945). Democracy in America (vol. 2; P. Bradley, ed.). New York: Alfred Knopf.

——— (1955). The Old Regime and the French Revolution (Doubleday Anchor Book ed.). Garden City, N.Y.: Doubleday.

TULLOCK, G. (1965). The politics of bureaucracy. Washington, D.C.: Public Affairs Press.

VILLANI, A. (1972). "La distribuzione di funzioni fra diversi livelli di governo." Città e Società, Milan, 7(March-April):3-34.

WOODCOCK, G. (1967). "Regional government: The Italian example." Public Administration, 45(winter):403-415.

——— (1969). "Aspects of the Italian regional system." Il Politico, Pavia, 34(March): 106-125.

Part III

CONSIDERING STRUCTURAL CHANGES

5

Local Government in Postindustrial Britain: Studies of the British Royal Commission on Local Government

ROGER W. BENJAMIN

☐ INSTITUTIONAL ARRANGEMENTS often lag behind the functions which they are designed to serve. Changes in material conditions and social relationships create pressures for institutional changes. In the societies which industrialized in the 19th century, problems of local government boundaries, the structure and functions of local authorities, and coordination between local, regional, and central governments were under scrutiny. But what changes should be undertaken today? It is easier to be critical than to offer constructive alternatives.

The study of local government presents problems representative of those found in all fields of political science. One needs data, but aggregate census and fiscal data are collected for purposes other than those intended by the researcher. Surveys are expensive, typically noncomparative, and not repeated. Conceptual issues require attention as well. Local government is concerned with the efficient provision of goods and services required by citizens. How do we define efficiency? What is meant by local government? What goods and services should be provided? What about the future of local government in postindustrial societies where citizen demands on governments appear to be unlimited? Is there an optimum balance between participation and authority? Finally, while there are no magic keys which will unlock solutions to these issues, the absence of theory must be identified as the major problem in the study of local government.[1] Without theory, contradictory conclusions and policy recommendations may be reached, sometimes from the same data. The results of ecological and survey studies cannot be evaluated properly because the interpreter does not have a structuring principle from which he may analyze the statistical findings. Such findings have meaning only if they corroborate or falsify hypotheses grounded in deductively generated theory (Brunner and Liepelt, 1972).

In this chapter I shall approach these questions through an anatomy of local government reorganization in England. In examining this major set of public policy changes, I shall draw some inferences about the relationship between research and reform in local government by applying lessons from the public choice approach (Ostrom and Ostrom, 1971) and from recent work in the study of comparative political change.

LOCAL GOVERNMENT REFORM

In 1972 local government in England was reorganized under the provisions of the Local Government Act (1972) to come into effect in 1974. The act and this analysis deal with local government in England excluding Scotland or Wales (Wheatly, 1969). The act stipulated considerable change in the existing local government structure which had been based on the Local Government Act of 1888. Although the reorganization effected does not wholly follow the recommendations of the Royal Commission on Local Government in England (1969), the arguments and conclusions of the Commission (the Redcliffe-Maud report) were largely accepted by the government as guides in their law-making and implementation efforts. These recommendations offer a useful point of departure for this chapter.

From 1966 to 1969 the Royal Commission examined written evidence from government departments and local government associations, received written and oral evidence from 2,156 witnesses, either as private individuals or representing professional organizations, carried out a substantial research program of its own, and commissioned a number of studies by independent institutions. Finally, the Commission had the benefit of numerous related commissions and government sponsored studies on welfare, the National Health Service, and local government itself. Perhaps there are government commissioned studies of comparable magnitude in public policy fields in the United States or elsewhere, but it is difficult to imagine a more exhaustive study of local government organizations.

THE PROPOSALS AND REFORMS ADOPTED

The Commission found the following structural weaknesses in English local government: (1) fragmentation of government units (79 county boroughs and 45 counties), (2) division of responsibility between upper and lower tiers within counties and between counties and county boroughs, (3) the small size of many authorities, which resulted in weakness in dealing with central government and difficulty in recruiting professional staff, (4) the arbitrary division between town and country, and (5) an unsatisfactory relationship between local authorities and the public.

Following these conclusions about local authorities, the Labour government white paper (*Reform of Local Government in England,* 1970) suggested that the number of unitary authorities should be reduced from 58 to 51 in number. The Conservative government white paper which followed (*Local Government in England,* 1972) argued that because of the competing claims of *efficiency,* which demanded larger units, and *democracy,* which demand smaller units, compromises had to be reached. Tasks were to be broken up between two tiers of authorities, counties and smaller districts. The allocation of functions was to differ between highly industrialized areas and rural areas. In conurbations (Standard Metropolitan Areas) effective organization of responsibility of important services such as education and personal social services were felt to require a population base of between 250,000 to one million and therefore it was possible to draw up a compact second-tier of metropolitan districts. These metropolitan districts were smaller than counties as the area-wide authority. In rural areas the second-tier authorities would have to be much smaller in population if they were to be reasonably compact in area. Here the main functions were to remain with the county council. Rural parishes were to continue while parish councils were to be formed in former boroughs and urban districts.

The Conservative government reorganization forced county

amalgamation and the merger of county boroughs with counties. At the county level the distinctions between town and county are eliminated. In size, all the metropolitan counties exceed one million, and the metropolitan districts average 100,000. Functions are divided between two tiers. The top tier has responsibility for functions which are felt to demand more uniformity or greater control by central government. This includes essentially all high capital investment in nonmetropolitan districts. Metropolitan districts maintain responsibility for education and other social services.

In sum, the 1972 Conservative reorganization follows the spirit and the assumptions of the 1969 Redcliffe-Maud report, thus marking another stage in the process of increasing the size of local government and centralizing the functions performed by it. The number of top-tier county authorities was reduced to less than 40% of the previous total, while the total number of local government units was reduced from just over 1,000 to slightly over 400. The total number of local councillors was reduced to 25,000; about 11,600 councillors and over 4,000 aldermen were eliminated. Central government retains a much greater control over local government than in the United States. Through the legal principle of *ultra vires,* local authorities are not allowed, by Parliament, powers that are not explicitly conferred upon them. Local councils must have explicit statutory authority for everything they do. Central government also controls local finance through grants and audits and by requiring local authorities to submit their proposals for borrowing to the central government. Government departments also supervise the work of local authorities, especially in education, police, and urban regional planning.

Questions of optimum local authority size and differentiation of functions between upper and lower tier authorities are related to the underlying issue of organizational efficiency. The new government reorganization incorporated the ideas of commissions which worked on related matters throughout the 1960s (Maud Committee, 1964; Mallaby Committee, 1967; Bains Committee, 1972). These commissions all endorse the principles of greater professionalization, meaning better trained staff, more staff to serve increasingly specialized functions, and the use of modern techniques of management as contributing to increased efficiency.

Next I shall criticize the Redcliffe-Maud Commission reports and studies.

AN INTERNAL CRITIQUE

The charge of the Commission was

to consider the structure of Local Government in England, outside Greater London, in relation to its existing functions; and to make recommendations for authorities and boundaries, and for functions and their division, having regard to the size and character of areas in which these can be most effectively exercised and the need to sustain a viable system of local democracy. [Royal Commission on Local Government, 1969, vol. 1, p. iii]

The Commission concluded that England should be divided into 58 unitary authorities plus 3 conurbations, where authority would be divided between an upper tier responsible for planning, transportation, and development functions and a number of lower tier metropolitan districts responsible for education, personal social services, health, and housing. The Commission further recommended that eight provincial authorities be set up with regional planning functions. In addition, local councils were to be encouraged so as to "represent and communicate the wishes of cities, towns, and villages" (Royal Commission on Local Government, 1969, vol. 1, p. 2). Moreover, I note the underlying question which governed Commission thinking:

What size of authority, or range of size, in terms of population and of area, is needed for the democratic and *efficient* (my emphasis) provision of particular services and for local government as a whole? [Vol. 1, p. 3]

The test of the validity of these conclusions and the assumptions upon which they are built rests in the appraisal of the Commission's evidence. Earlier I noted the extensive evidence secured by the Commission. The evidence from the government departments not surprisingly supported the unitary principle, which argues for larger units and reduced fragmentation of authority. The arguments run parallel to those assumed by the Commission to be correct. Economies of scale would bring the benefits of more and better trained professional staff; it would reduce costs and facilitate planning in a number of ways; "Government departments left us with the impression that, were it not for democratic considerations, they would really like a system of 30 to 40 all-purpose authorities"

(Vol. 1, p. 43). The rationalization of linkages between levels of government would lead to greater efficiency of local government operation.

INTERPRETATION OF THE EVIDENCE

The evidence from the research program of the Commission presents the reader with puzzling contradictions. The heart of the research program dealt with the relationship between size and performance. In language perhaps unacceptable to social scientists it is concluded (Royal Commission on Local Government, 1969, vol. 1, p. 58) that,

> the over-riding impression which emerges from the three studies by outside bodies and from our own study of staffing is that size cannot statistically be proved to have a very important effect on performance.

Yet the report goes on to conclude from this that, "since all the statistics used were necessarily compiled on an existing local authority basis, they could not tell us how a new pattern of authorities might perform" (vol. 1, p. 58). The new structures would differ from the old so we cannot infer that the same lack of relationship between size and performance would be found to exist in the newly constituted authorities. This extraordinary conclusion suggests the difficulties yet to be overcome in public policy relevant research. The only thing correct about this portion of the Commission's position on the size-performance relationship is that it is unwise to suggest that findings from one study can be assumed to hold in a possibly divergent setting. However, all that empirical tests of relationships (hypotheses) can ever hope to do is to corroborate hypothesized relationships which stem from prior expectations (a theory or theory notion) about the way variables are related, or falsify the hypothesized relationship by disconfirming it. The Commission members should have evaluated the reliability and validity of the specific research studies, of course; but to reject a series of statistically significant findings which all run in the same direction, namely that size and performance are not related positively, points to the strength of a priori images and assumptions in the minds of Commission members which led them to reject nonsupportive evidence. Faced with disconfirming evidence from the statistical studies, the Commission simply relied more heavily on the impressions of central government departments and thus went on to reaffirm the assumptions they began with. Greater size of local authorities will bring better and more specialized staff. "They also make it possible to achieve a more rational distribution of staff and institutions. Larger administration has the additional advantage of spreading the administrative load more evenly" (vol. 1, p. 59). In the absence of another set of assumptions about the relationship between size and performance in local government, based on a competing theory, it appears as if the Commission groped to affirm the unitary principle whenever and wherever possible.

Two other general conclusions of the Commission which are of interest concern democracy and financial structure. It would be disingenuous to accuse members of the Commission of an anti-democracy bias. The report suggests that they accepted the recommendation of the need for increased participation in local government suggested by the Skeffington report (Skeffington, 1969), and they authorized a full-scale community survey of citizen attitudes toward local authorities (Research Services Limited, 1969). The first difficulty, one recognized by the Commission, is conceptual, i.e., what are meant by democracy, participation, and community? The second problem is how to develop operational measures on the domains of related concepts. Community: what constitutes community in terms relevant for the construction of appropriate local government units? Local government accessibility and responsiveness: to what extent do citizens perceive local government authorities as accessible to them and responsive to their demands? Decentralization: what is the relevance of local councils for greater participation? Representation and community attitudes: to what extent do the individual characteristics of the electors and the elected, e.g., sex, age, length of residence, education, and their respective attitudes toward local government mirror one another and what are citizen attitudes toward concepts of community in terms of alternative sizes and types of local authority structures?

The community survey found that most citizens tend to identify with their local community as their home area, no matter what the population size. Though not commented upon directly, the high voter apathy toward local government concerned the Commission members as well. When asked whether they preferred a reformed system of local government to the present one, most citizens said they desired the status quo. The Commission chose to interpret this as irrelevant because, from their viewpoint, most people will indicate a preference for the status quo unless there is very substantial evidence before them indicating the need for change. This may be so, but why? Another view, based on the public choice approach to be introduced below, is thus: citizens will forgo governmental (or any organizational) change affecting them unless the net benefits of the organizational change are perceived as outweighing the net benefits of remaining with the status quo. And, not surprisingly, elected officials are not representative of the citizens they serve in terms of individual characteristics; they tend to be male, older, and better educated than the citizens they serve. No relationship was found between interest in local government and the size of the local authority; apathy toward local government, measured by name recognition of councillors, is apparently distributed equally throughout the local authorities regardless of size (Research Services Limited, 1969).

The difficulty with financial structure concerns the rise in expenditures. The Commission contented itself with noting that the steady rise in the proportion of gross national product devoted to local government (about 15% in 1966-1967) must decline at some point in the future (Royal Commission on Local Government, 1969, vol. 1, p. 130). One Commission member stated that local government expenditures may rise to 24% by the turn of the century (Senior, 1969:148). This seems to me an extraordinary projection given the state of the British economy, even in 1969. A reading of the local government financing projections (Royal Commission on Local Government, 1969, vol. 3, pp. 95-128) suggests how difficult it would be to keep local government finance on a level with inflation—and all this before the double-digit inflation of the 1970s. Nowhere does one find discussion of how to handle the question of priorities among the social services which local authorities would face when financial growth slowed or stopped (Lapping, 1970). Neither did the Commission recognize that greater professionalization, itself, is likely to result in greater costs. Professionalization appears to be associated with more pay, less work, and increasingly liberal retirement programs. This consequence is a function of the stronger bargaining accruing to those who enjoy a constrained and selective market situation.

So, the Commission recommended unitary authorities which would remove the artificial boundaries between town and country and a minimum "ideal" size of authority of around 250,000, with 1,000,000 population catchments cited as the upward limit, though in the words of the report (vol. 1, p. 71):

Our own conclusion is that there is no single service in which administration by a very large authority would have decisive disadvantages. Future developments in most services seem almost certain to favor much bigger operational units than most of the existing ones.[2]

THE SENIOR DISSENT

Before turning to an external critique of local government reform in England, I wish to comment on the Memorandum of Dissent written by a Commission member, Derek Senior (1969) which elaborates on several of the criticisms presented here. He presents a criticism of the unitary principle, notes the propensity of bureaucrats to select larger minimum size population units than is necessary, and supports the idea of a local council (he defines it as a "common" council) with political powers rather than one which merely serves a representational function. He correctly interprets the statistical studies not only as falsifying the hypothesis that economies of scale are associated with size but, in several cases, as indicating diseconomies of scale. Instead of a unitary local authority system, Senior recommends a two-tier system organized around a functionally defined (in terms of spatial boundaries isolated by relative densities of population) city-region set of counties. Senior envisages the "common" council only where the sense of community is strong enough to demand one.

In one sense, Senior's strategy is superior to the Commission recommendation which prescribes rigidly that there must be local councils in every borough or county district. However, balanced against the self-organization principle are a series of questions related to resource distribution for the quality of life of both rural and urban inhabitants. D.M. Hill (1973) argues, for example, that local government boundaries may act to perpetuate socioeconomic inequalities, inequalities which do exist in England (Royal Commission on Inequalities of Wealth, 1975). Local authorities that organize to develop neighborhood parks, libraries, and other social services may draw a greater than average share of the metropolitan area's resources. Whether, to what extent, and how local governmental units should be used to redress socioeconomic inequalities is a complex problem involving conceptualizations of the ideal polity, social justice, and the effects of various proposed reforms.

Why were Senior's criticisms of the unitary principle and suggestions ignored? One suspects it was because the Senior proposal was not built on a clear theoretical base which would commend itself strongly enough in comparison to the unitary concept so tenaciously held to by his colleagues. The next section takes up this problem.

PUBLIC CHOICE AND LOCAL GOVERNMENT REFORM

The Commission recommendations betray confusion over the goals of local government and the theory (or theories) about how it works or, more important, how it might be changed to work better. How might we improve this situation?

Two alternative structuring principles are provided by the theory of public goods (Mishan, 1972) and recent work in comparative political change. Briefly the relevant assumptions of the theory of public goods necessary for the model proposed are the following. First, the theory employs the familiar assumption of rational choice on the part of the individual. I assume that individuals act in such a way as to maximize gains or minimize losses; they will choose alternative A when its net benefits are perceived to outweigh alternative B's net benefits. We may attach to this simplifying assumption about individual motivation a number of nonobvious aggregation rules about collective action, i.e., the conditions under which individuals will join in group action, comply with leadership wishes, etc. (Olson, 1965). Second, I shall adopt the view of government as the provisioner of goods, some "pure" public and some "mixed" with spillover effects on citizens which are positive or negative externalities. Analytically, pure public goods (or bads) are those goods and services provided by the government whose benefits (or costs) may not be withheld from any citizen of the collectivity in question. Examples include national defense, police protection, air pollution control. When provided, all citizens enjoy their benefits; hence the opportunity for particular individuals to enjoy benefits without paying their share of the costs (the free-rider problem) also exists. Conceptually polar to public goods are private goods which may be purchased and consumed only by the individual, e.g., food, housing, clothes. Two points remain to be clarified. In between pure public and private goods are a wide range of goods which possess the quality of jointness; their benefits and costs are not shared equally but are distributed differentially throughout the population. From the production of goods in factories we receive negative externalities like pollution. From government we receive goods such as roads which may have unintended negative spillover effects for the communities adjoining the roads, such as noise pollution and the destruction of existing neighborhoods. These goods are often referred to as collective goods by theorists because the logic applies to any jointly consumed goods, whether or not they are produced by government (Samuelson, 1967:47). I argue below that local government in postindustrial society is increasingly about conflict generated by spillover effects, free-rider problems, and negative externalities created by the continuing growth of government in societies which are reaching new stages of interdependence (LaPorte, 1975).

The implications of the theory of public goods for thinking about alternative institutional structures for English local government are significant. The approach offers a different way to think about the unitary principle and associated economies or diseconomies of scale. Though even central ministries indicated ambivalence about the impact of ever increasing centralization of government functions in large public bureaucracies, none of the Commission members or witnesses appear to have before them the public choice structuring principle which would have allowed them to look at local government in a different light. Let us again look at the two basic problems considered by the Commission.

SIZE AND EFFICIENCY OF LOCAL AUTHORITIES

The Commission members, with the exception of Senior, accepted the view that size and efficiency are correlated in a linear manner, the greater the size, the greater the level of efficiency. Of course the Commission is in good company; this view is dominant among students and practitioners of local government (compare V. Ostrom, 1974). Greater size is assumed to lead to the achievement of a "critical mass," a point defined as the existence of an organization sufficiently large to allow enough specialization necessary to perform the function relevant to its organizational goals competently. It is felt also that greater size promotes economies of scale. One single-purpose educational bureaucracy can purchase equipment at a discount compared to smaller educational authorities. However, the public choice approach suggests a different view of the likely effects of size on efficiency,[3] which alerts us to the probability that, beyond a certain size, inefficiencies are likely. This hypothesis is based on the following argument, which benefits from Tullock (1965) and Downs (1967).

Bureaucratic organizations may be perceived as structures staffed with self-interested, rational, goal-seeking individuals. Individual organization members are primarily interested in careerist objectives. In theory, advancement up the organization accrues to those who perform assigned tasks most successfully. In fact, however, rewards are most likely to go to those who perform functions in a manner most congruent with the organization's latent goals. These include selection of non-risk-taking problem solutions so as not to risk failure and, most importantly, the transfer of information which inferiors feel their superiors will want to hear. Information which indicates bureaucratic failure is unwelcome because it requires action which may have uncertain consequences for the superior. All of this results in information bias. The more hierarchical the levels that information passes through, the greater the possibility for information to be biased. Further, the same information bias occurs between departments which, after all, compete for funds, power, and advantage within the larger organizational system. When we look at the relationship between size and efficiency we should therefore expect to find diseconomies, not economies, of scale to accrue beyond a certain size.[4]

This different theory perspective gives us an alternative set of expectations concerning the research findings of the Commission. The Commission studies do not lend support to the expected positive relationship between size and efficiency, but inspection of the two best designed studies shows support for the public-choice-based interpretation. These studies, relying on econometric techniques, corroborate the public choice hypothesis about size and efficiency. For example, with respect to housing, the York study concludes,

> Considering "cost per foot" as the relevant dependent variable, our findings for the County Boroughs suggest the opposite of what has been the common view about the effect of an increase in the population or "size" of a local authority on its efficiency. . . . Diseconomies of scale operate with an increase in the population size. [Gupta and Hutton, 1968:6]

With respect to another measure, management costs per dwelling unit, the study concludes that for noncounty boroughs and rural district councils the hypothesis concerning economies of scale are refuted. And for urban district councils, our findings suggest a u-shaped supervision and management cost curve with relation to population, such costs being kept at their minimum when the population size of an Urban District Council is about 40,000 beyond which tending to rise. Therefore, an increase in their size beyond 40,000 would also give rise to diseconomies of scale. [Gupta and Hutton, 1968:7]

The York study authors find the same u-shaped curve for cost and population for health care measures and another inverse curve for size of population of authorities and the expenditure per mile of highway construction.

The second study examines the relationship between performance and size of local educational authorities. The authors of this study, also using regression analysis of expenditure data, find that larger authorities spend more, not less, than small authorities on texts (Local Government Operational Research Unit, 1968:40) and that the cost of maintenance and replacement of educational equipment increases sharply with the size of the population being served by the educational authority. In addition, larger authorities appear to have fewer, rather than proportionately more, specialist advisors than their smaller counterparts; there is an increase in expenditures for specialists and advisors from the very small to medium-sized authorities but a decline in expenditures after that point as one moves to the larger authorities (Local Government Operational Research Unit, 1968:112).

THE NEED TO DIFFERENTIATE AMONG PUBLIC GOODS

Most Commission members distinguished between the size and type of government required by intensive face-to-face social services versus those more appropriate for larger administrative units such as water or sewage services. A return to the public-private good discussion suggests a different view concerning the question of the governmental organization appropriate for the delivery of different types of public goods. Though research on the implications of the nature of public goods for the optimum size and type of govern-mental units is only beginning, several points are already clear. Where services are more rather than less information sensitive, smaller not larger government units are to be preferred. While, for example, the errors associated with the information transfers connected with the problem of sewage disposal may not be large, this is unlikely to be the case with educational, social, welfare, or health services. Next, free-rider problems tend to occur where user charges are not levied and where there are few negative sanctions or positive incentives for the individual to pay. This can mean, as appears to be the case with the British National Health Service, that underinvestment in capitalization areas occurs, coupled with overutilization of the services since they are "free." The result is a health service which is only sporadically shored up financially by central government. Alternatively, small government units may create islands of privilege by developing high quality services for their residents while excluding nonresidents from using those services even though they may be paying for them through national taxes. It may also be the case that nonresidents may gain significant free-rider benefits by being able to enjoy the benefits of neighborhood parks, concerts, and other locally provided programs without paying their share of costs which fall on residents of the authority. This may result in greater costs than necessary for some governmental units.

Finally, the need for many specific types of social services are sporadic. For instance, the need for organizations in health care and police protection large enough to support specialists able to handle any contingency is cited as an argument for centralization. But what do the riot squads or special kidney transplant teams do when they are not being used? Often smaller government units which operate with part-time or volunteer support provide more flexible and less costly service (E. Ostrom et al., 1972). Perhaps too much specialization too well organized is too much of a good thing after all. Should one be surprised at the research findings of diseconomies of scale related to increased size of government units? It seems rather that economies of scale may be achieved when dealing with more mechanized services such as air pollution control and transportation networks, while diseconomies of scale and u-shaped curves flow from other services. This should alert us to the need for differentiating public goods. Moreover, another structuring principle, from recent work in comparative political change, suggests the same point for different reasons.

THE IMPACT OF POSTINDUSTRIALIZATION

The study of comparative political change has moved from earlier work that was biased toward democratic and linear notions of change to the characteristics of change itself. However the welcome focus on change is not a substitute for theory nor a substitute for rethinking the basic foundations of the political change process, the paradigm within which the investigator operates. The point is that we have a new benchmark, the concept of postindustrial society (Inglehart, 1971; Bell, 1973; Huntington, 1974; Ike, 1973). It is not too early in the study of the subject to remind ourselves not to fall into the conceptual, assumptional, and methodological traps that consumed much thinking about the political implications of modernization and industrialization. It is important to continue to refine our emphasis on change and to not use the postindustrial concept statically.

If industrialization is about change, growth, and the construction of basic political institutions, postindustrialization is the period in which it is time to deal with the impact of those changes—to shift from the emphasis on growth to considerations of equity and to modify or redesign political institutions which are no longer congruent with the needs and expectations of citizens. It may be sufficient to identify the central indicators of the process of change leading from industrialization to postindustrialization.

In the economy the change may be conceptualized as the point at which the shift in relative growth from the manufacturing or industrial sector of society to the service, including the public, sector begins. It is the point when the emphasis shifts from the production of steel and basic manufactured goods to what Bell (1972) sees as the codification of knowledge itself, the process of research and development. Societies move into this period at different points and with differing emphases. The American economy began this shift in the 1940s or earlier, and Britain has been in the same process of change from the 1950s (Benjamin, 1975). In Britain, as in the United States, the manufacturing sector of the economy has declined, first relatively and recently in absolute terms, compared to the service and especially the public sector of the economy. In Britain, especially, the decline of the private sector accompanied by the continued rise of the public sector has resulted in a declining tax base at a time when it is greatly needed. The overall problem is that the yearly productivity increases in the private sector of the economy have been used until now to set the tone for wage increases for all sectors of the economy. This becomes less satisfactory as the economy moves into a slower real growth cycle as measured by the traditional gross national product measures. Baumol (1967:109) puts the argument for costs in government thus:

> Economic activity can be divided into two types: technical progressive activity in which innovating capital accumulation and economies of large scale all make for a cumulative rise in output per man-hour and activities which, by their very nature, permit only sporadic increases in productivity. If productivity per man-hour rises cumulatively in one sector relative to its rate of growth elsewhere in the economy, while wages rise commensurately in all areas, then relative costs in the non-progressive sectors must inevitably rise, and these costs will rise cumulatively and without limit.

In addition to evidence concerning the growth in negative externalities associated with the legacy of industrialization, there are data to suggest that one major economic feature of postindustrial society is the concern for equity in the distribution of wealth, not its growth. Though research on this problem is just starting, it may be that, after a minimum welfare floor level is achieved in industrialized societies, economic inequalities between classes do not decline (Royal Commission on Inequalities of Wealth, 1975). Hirschman (1973) argues that if economic growth is not followed by substantial redistribution of wealth, class conflict will become heightened. This conflict may be deferred in a relatively homogeneous society because like citizens will wait in the expectation that they will eventually share in economic growth. However, when they perceive that this expectation is not likely to be fulfilled, the response by way of strikes, demonstrations, or other forms of political demands will be even more intense. England, the oldest industrialized society, has had the greatest amount of time to generate economic growth and achieve economic equality. If it is correct that socioeconomic inequality has remained relatively constant for decades (Royal Commission on Inequalities of Wealth, 1975), working class militancy and high demand levels on government find a possible explanation in the Hirschman model.[5]

There are studies to draw on with respect to the social and cultural change linked with postindustrialization following Inglehart's (1971) paper which showed that since 1945 younger people appear to be increasingly freed of material acquisition needs and are adopting "post bourgeois" values of intellectual and aesthetic need and "belongingness." This is true of the middle and upper classes, while workers' values, especially those of older workers, remain dominated by the need to protect their material and physical security. In politics the value change is conceptualized by Inglehart to result in preferences for participation in decision making and protection of freedom of speech versus bourgeois preferences for the maintenance of order and protection against inflation. The preference for political participation over authority maintenance finds echoes in studies of English local government, the Commission report itself, and cross-national studies. The question is why and what are the consequences of this for local government?

Industrialization in England brought many benefits. One might list a roll call of domestic increases in the quality of life as well as cite international triumphs. However, after a period of time spillover effects of industrialization began to mount. A centralized government sufficient to govern in the industrialized era becomes over-institutionalized (Kesselman, 1970) in the postindustrial period. Its large-scale bureaucracies become increasingly ill adapted to meet the growing and differentiated needs of the publics they serve. The growth of citizen demand is brought on by spillover effects throughout society which require public solutions plus the change in preference schedules of individuals from acceptance of the mere existence of services to the demand for increased quality of performance and eradication of existing inequalities (Brittan, 1975). *The change in the content of preference schedules may be assumed to be a rational response to fundamental changes in the socio-economic level of development.*

In England, as in other Western European societies, alternative educational groups desire basic reforms in the nature, not the provision, of education. Neighborhood community groups organize to fight negative externalities such as noise and air pollution or new highway construction and attempt to create higher quality local environments. These are collective action units spurred on by the environment in which they operate.

This sketch is simply an outline of a plausible explanation for the growth in participation demands in England; the research remains to be done. In retrospect we should not be surprised that the kinds of government appropriate for one era might need adjustment, reemphasis, or wholesale change in another time period. When flexible responses are required to meet diverse wants the previous highly centralized political institutions are unlikely to be appropriate governmental units if public-choice-based notions about the relationship between size and efficiency are correct. An increase in public goods and externalities can add up to increased divisiveness and conflict because diverse wants or values with respect to a collective [public] good are a basis for conflict, whereas different wants with respect to individual or private goods are not. . . . Everyone in the domain of a given collective good must put up with about the same level and type of collective goods, whereas with different tastes for private goods each individual can consume whatever mix of goods he prefers. [Olson, 1965:173]

If we return to the Commission report, the horatory dialogue about local participation provides support for my explanatory sketch. Assuming that further socioeconomic change in England will continue in the postindustrial direction, additional danger signals for the Commission recommendations are found in recent studies. In a follow-up to Inglehart's (1971) study, Marsh (1975) presents evidence of the continuing increase in the percentage of English who hold post-bourgeois values. Marsh divided his sample into two groups, acquisitive (materialist) and postbourgeois. Whereas 9% of the acquisitive group indicated dissatisfaction with the district in which they lived, 21% of the postbourgeois group expressed discontent (Marsh, 1975:25). In addition, the postbourgeois group were dissatisfied with the "unresponsiveness of elected representatives toward the electorate and more dissatisfied with the level of democracy in Britain" (Marsh, 1975:25). While the total economy has been growing slowly or experiencing negative growth, the local government share of gross national product has risen from 13.8% in 1964 to 15.4% in 1974. The interest and debt payment burden on local government has increased greatly in the past few years, and yet demands on government, at all levels, appears inexhaustible (King, 1975).

CONCLUSION

Compared to the United States (Banfield, 1960), pressure on local government in England has not been great. Only recently, for example, has the number of automobiles reached a level at or beyond the capacity of the road network. Social welfare case loads do not approach American levels and crime rates, though rising, are also small in comparison to those in the United States. Yet pressure on local government has begun to grow rapidly in recent years. Throughout the 1960s the proportion of Britain's gross national product devoted to the public service sector increased enormously.

While local government autonomy in England was small previously, the reorganization efforts in the past few years have resulted in even greater centralization. This in spite of counter research evidence discussed here. I turn to the implications of the argument.

(1) Finance. The commission mandate did not include consideration of local government finance. This was a serious omission. Superficially, English local authorities such as London (run by the Greater London Council) appear to be in better financial condition than, say, New York City, but this is appearance not reality. Most American local governments are limited statutorily from spending beyond set limits. Most of the operating revenues for English local authorities come from central government so there is no immediate danger of "bankruptcy." However, in England, as in other post-industrial societies, inflation, coupled with the slowdown in the rate increase in the gross national product, is likely to result in substantial deterioration of the entire public sector in the next decade in terms of its financial base (Lapping, 1970).

(2) Political participation. I have argued that the rise in participation demands is not a time-honored but meaningless plea for more democracy. It stems from two theory perspectives which are converging rapidly. First, the rise in spillover effects may tend to reinforce existing levels of economic and social inequalities through the failure to ameliorate those problems. These spillovers occur not only because of the effects of private industry but because government is so heavily involved in providing essentially "mixed" goods which, in turn, creates the need for local responses. Second, as individuals become better educated, have more leisure time, and are more free from material needs, there is evidence that their preference schedules change. Individuals and their collectives unite to demand differential treatment from large anonymous bureaucracies which find it difficult to respond to particularized preferences of smaller subsets of citizens. The growth of diverse wants may, in fact, be what postindustrialization is all about.

The future is likely to bring unpleasant surprises to adherents of the unitary principle of governance. "Hidden hand" modeling by operations research analysts conceal biases as broad and skewed as any other style of planning (Downs, 1969). There is no way I know of to correctly gauge the preferences of individuals without allowing them an opportunity to express them. Cost-benefit calculations may be made concerning the consequences of many issues, but this fails to take into account the diversity and intensity with which various preferences are held. Here the suggestions by Vincent Ostrom (1972) for institutional redesign are useful.[6] Ostrom's concept of poly-centricity, summarized by Sproul-Jones (1972), captures the need for autonomous and overlapping catchments in postindustrial societies,

> Legal authority is fragmented among federal, state and, in many cases, local sets of institutions, themselves fragmented into executive, legislative and judicial departments. None of these enjoys independence in the making of particular decisions. An individual who finds a decision made in one center unacceptable to his interests has the constitutional capacity to take his case to an alternative forum. The combined effect of these multiple decision structures, all of which are potent but non-omnipotent centers of power is to require the support for any public policy or choice of more than a mere majority of interested citizens. Any disaffected party to an economic transaction can exercise a veto-like position, for he can wield a potential threat of involving the other parties in protracted and risky political ventures. These potential political costs create a substantial incentive, therefore, for a mutual accommodation. . . . A similar incentive exists for those charged with legal responsibilities for the operation of public enterprises. . . . The public entrepreneurs have an incentive to come to terms with their public or clientele, and with other agencies and public officials jointly serving the clientele. Even though the behavior of the public enterprises may, as a consequence, appear imperialistic, . . . such behavior need not, in short, be seen as pathological, but as a type of public analogue of private competitive entrepreneurial behavior. [Sproul-Jones, 1972:185-186]

SUGGESTIONS

The principles upon which I would proceed would be as follows. First, it is important to design local authorities and public bureau-cracies to meet single-purpose and multipurpose functions within a representative community base which is defined along the lines suggested by the Senior dissent (1969). In doing so, principles emanating from the criticisms presented above should be imple-mented. Since public goods delivered by government differ, so should the government units designed to deliver them. A legal authority spread over a wide geographic area and cutting horizontally across many government units may be appropriate for dealing with the Thames tidal basin. Small government units may, however, be more appropriate for personal social services which, with their labor intensive aspect and concomitant greater propensity for information distortion in larger bureaucracies, appear more suitable for implementing the imperatives of the Royal Commission on Local Government to coordinate personal social services to the needs of the individual and the family. In this respect the two-tier division of responsibilities makes matters worse. No doubt the two-tier notion is essentially a principle developed in response to the recognition of the differences in types of public goods. However, the power-sharing arrangements discussed (Jones, 1973) only create more uncertainty and delay arising from communication difficulties and misunderstanding over responsibilities. A better strategy would be to give local authorities and particular public corporations maximum legal powers consistent with their being able to carry out their functions. This would include taxation powers whenever possible.

Postindustrial society brings new problems to local government, and the fundamental alteration in their design and in the demands placed upon them is not far off. The reasons that unitary based local government *worked* in the past were probably never the reasons thought to be the case anyway. The low burdens on English local government were surely a measure of the "successful" governance of local communities until the past decade. In any case, in the 1970s the problems encountered in English urban and rural life begin to sound distressingly similar to North America's big cities: rising crime rates, minority problems, traffic and pollution issues. The unitary approach to local government, which has enjoyed higher currency in England because of the historical propensity of citizens to allow elites wide latitude in governance (Schwartz and Wade, 1972), has been questioned in this chapter. Perhaps it is time to begin to think in terms of political ecology rather than the political development of local government, just as ecology—the qualitative management of finite human and economic resources—is becoming the substitute for growth.

NOTES

1. In fact, Ashford (1975) finds that local government expenditure patterns controlled by national government are very similar in Britain and the United States. However, I feel that the factors I have listed suggest that much greater central control of local government exists in England than in the United States.

2. Note, further, the following associated conclusion: "in services where size of population is especially important, once an authority's population goes much above 1,000,000 further gains in functional efficiency are unlikely to offset disadvantages associated with the management of such large units. . . . The concentration of work in a single authority in charge of all local government services could be too great if the authority was responsible both for an extensive area and for a population of well over a million. It could have serious managerial problems due to the sheer size and complexity of the organization it would have to maintain. Perhaps they could be overcome but much continuing effort would have to go into solving them" (Royal Commission on Local Government, 1969, vol. 1, p. 71). Why are these problems considered relevant for bureaucratic organizations serving areas in excess of one million? Why not one-half million, or even less, or, for that matter, what grounds does the Commission have for saying there would be any additional problems in a population area of over one million?

3. To my knowledge the concept of efficiency is nowhere defined by those writing about English local government. It is a difficult problem; is the police force that arrests more individuals more successful than the one which arrests fewer but perhaps this is due to crime prevention methods? One way to view the concept is in terms of the organization's *success* in making choices which yield a higher proportion of benefits over cost vis-à-vis its stated organizational goals.

4. The above argument lists basic negative conditions leading to information bias and diseconomies of scale. Johnson (1975) lists representative positive conditions which contribute to bureaucratic growth for its own sake. Growth of the size of the organization enhances the power of the organization leaders. It also is a generally accepted measure of the management competence of the leaders. Growth is popular with existing staff because it creates new opportunities for advancement; this leads, at a minimum, to the sacrosanctity of existing budgetary allocations *and* the toleration of wide margins of inefficiency.

5. It is important to note two additional points which Vincent Ostrom called to my attention. First, absolute social and economic equality is probably not only impossible but undesirable for reasons which I shall not list here because of the length needed to cover such a topic. Second, postindustrialization may actually increase the number of poor in all remaining unorganized groups in society, e.g., the young, the aged, women, and minorities.

6. For some (Self, 1975) Ostrom's concept of policentricity offers an apolgia for the fragmented world of American local government. In fact, it is the start of needed design work to develop a new structuring principle from which local government reform may proceed.

REFERENCES

ASHFORD, D.A. (1975). "Parties and participation in British local government and some American parallels." Urban Affairs Quarterly, 11(1):58-81.

BANFIELD, E.C. (1960). "The management of metropolitan conflict." Daedalus, 90(winter):61-78.

Bains Committee (1972). The new local authorities: Management and structure. London: Her Majesty's Stationery Office.

BAUMOL, W.J. (1967). "Macroeconomics of unbalanced growth: The anatomy of urban crisis." American Economic Review, 57(June):415-426.

BELL, D. (1973). The coming of post-industrial society. New York: Basic Books.

BENJAMIN, R.W. (1975). "The political consequences of postindustrialization." International Review of Community Development, 33-34(winter):149-158.

BRITTAN, S. (1975). "The economic contradictions of democracy." British Journal of Political Science, 5(2):129-159.

BRUNNER, R.D., and LIEPELT, K. (1972). "Data analysis, process analysis, and system change." Midwest Journal of Political Science, 16(4):538-569.

DOWNS, A. (1967). Inside bureaucracy. Boston: Little, Brown.

——— (1969). "The coming revolution in city planning." Pp. 596-610 in E.C. Banfield (ed.), Urban governments. New York: Free Press.

GUPTA, S.P., and HUTTON, J.P. (1968). Economies of scale in local government services (Research study 3, Institute of Social and Economic Research, University of York). London: Her Majesty's Stationery Office.

HILL, D.M. (1970). Participating in local affairs. London: Penguin.

HILL, R.C. (1974). "Separate and unequal: Governmental inequality in the metropolis." American Political Science Review, 68(4):1557-1568.

HIRSCHMAN, A.O. (1973). "The changing tolerance for economic inequalities." Quarterly Journal of Economics, 87(4):544-566.

HUNTINGTON, S. (1974). "Postindustrial politics: How benign will it be?" Comparative Politics, 6(1):152-169.

IKE, N. (1973). "Economic growth and intergenerational change in Japan." American Political Science Review, 67(4):1194-1203.

INGLEHART, R. (1971). "The silent revolution in Europe: Intergenerational change in post-industrial Europe." American Political Science Review, 65(4):991-1017.

Institute of Local Government Studies, University of Birmingham (1968a). Administration in a large local authority: A comparison with other county boroughs (Research study 10). London: Her Majesty's Stationery Office.

——— (1968b). Aspects of administration in a large local authority (Research study 7). London: Her Majesty's Stationery Office.

JOHNSON, H.G. (1975). On economics and society. Chicago: University of Chicago Press.

JONES, G.W. (1973). "The Local Government Act 1972 and the Redcliffe-Maud Commission." Political Quarterly, 44(2):154-166.

KESSELMAN, M. (1970). "Overinstitutionalization and political constraint: The case of France." Comparative Politics, 3(3):21-45.

KING, A. (1975). "Overload: Problems of governing in the 1970's." Political Studies, 23(2):162-174.

LAPORTE, T. (ed., 1975). Organized social complexity: Challenge to politics and policy. Princeton, N.J.: Princeton University Press.

LAPPING, B. (1970). "Which social services can we save?" Pp. 180-193 in W.A. Robson and B. Crick (eds.), The future of the social services. London: Penguin.

Local Government Act, 1972 (1972). London: Her Majesty's Stationery Office.

Local Government in England (1972). London: Her Majesty's Stationery Office.

Local Government Operational Research Unit (1968). Performance and size of local education authorities (Research study 5). London: Her Majesty's Stationery Office.

Mallaby Committee (1967). The staffing of local government. London: Her Majesty's Stationery Office.

MARSH, A. (1975). "The 'Silent Revolution,' value priorities, and the quality of life in Britain." American Political Science Review, 69(1):21-30.

Maud Committee (1967). The management of local government. London: Her Majesty's Stationery Office.

MISHAN, E.J. (1972). Elements of cost-benefit analysis. London: George Allen and Unwin.

OLSON, M. (1965). The logic of collective action. Cambridge, Mass.: Harvard University Press.

OSTROM, E., PARKS, R.B., and WHITAKER, G.P. (1973). "Do we really want to consolidate urban police forces? A reappraisal of some old assertions." Public Administration Review, 33(September/October):423-433.

OSTROM, V. (1972). "Polycentricity." Paper delivered at the annual meeting of the American Political Science Association, Washington, D.C., September.

――― (ed., 1974). "The study of federalism at work." Publius, 4(fall).

OSTROM, V., and OSTROM, E. (1971). "Public choice: A different approach to the study of public administration." Public Administration Review, 31(March/April):203-216.

OSTROM, V., TIEBOUT, C.M., and WARREN, R. (1961). "The organization of government in metropolitan areas: A theoretical inquiry." American Political Science Review, 60(4):831-842.

Reform of Local Government in England (1970). London: Her Majesty's Stationery Office.

Research Services Limited (1969). Community attitudes survey: England (Research study 9). London: Her Majesty's Stationery Office.

RICHARDS, P.C. (1973). The reformed local government system. London: George Allen and Unwin.

Royal Commission on the Distribution of Income and Wealth (1975). Report. London: Her Majesty's Stationery Office.

Royal Commission on Local Government (1969). Report of the Redcliffe-Maud Committee (3 vols.). London: Her Majesty's Stationery Office.

SAMUELSON, P.A. (1967). "Indeterminancy of governmental role in public-good theory." Papers in Non-Market Decision Making, 3:47.

SCHWARTZ, B., and WADE, H.W.R. (1972). Legal control of government: Administrative law in Britain and the United States. Oxford: Clarendon Press.

SELF, P. (1975). "Economic ideas and government operations." Political Studies, 23(2-3):359-389.

SENIOR, D. (1969). Memorandum of dissent (Report of the Royal Commission on Local Government, vol. 2). London: Her Majesty's Stationery Office.

SKEFFINGTON, A. (1969). People and planning. London: Her Majesty's Stationery Office.

SPROULE-JONES, M. (1972). "Strategic tensions in the scale of political analysis." British Journal of Political Science, 2(2):173-191.

TULLOCK, G. (1965). The politics of bureaucracy. Washington, D.C.: Public Affairs Press.

Wheatly Commission (1969). The reform of local government in Scotland. London: Her Majesty's Stationery Office.

WOOLF, M. (1968). Local authority services and the characteristics of administrative areas (Research study 5). London: Her Majesty's Stationery Office.

6

Restructuring Service Delivery Systems in West Germany

ARTHUR B. GUNLICKS

☐ THE MID-1960s MARK THE BEGINNING of a decade of far-reaching structural, boundary, and administrative reforms in local government in West Germany. During the past decade the eight "territorial states" *(Flächenstaaten)*[1] have recast about 24,500 cities, towns, and villages *(Gemeinden)*—which we shall call "municipal-ities" in this paper—into fewer than 11,000 units within a variety of structural forms. Of the 425 counties *(Landkreise)*, only 250 remained by July 1, 1975 (Buchsbaum, 1975:407; Wimmer, 1975:464). These numbers will be reduced even further when Lower Saxony completes its county reforms and Bavaria its municipal reforms within the next year or two.[2]

The purpose of this paper is (1) to review the basic arguments used in support of these structural, boundary, and administrative reforms, (2) to present some of the general standards used in planning and carrying out the reforms, (3) to note the similarities and differences among the eight reforming states in the implementation of structural and administrative changes, and (4) to consider opportunities for

AUTHOR'S NOTE: *I wish to thank the German Fulbright Commission and the University of Richmond for making it possible for me to spend a sabbatical year in the Federal Republic of Germany.*

[173]

assessing the general impact of the changes—since the reform efforts are based upon presumptions that certain changes will yield improvements in the performance of the newly arranged units and structures of local government.

RATIONALE FOR REFORM

CONSTITUTIONAL ARGUMENTS

The 1949 Constitution or Basic Law *(Grundgesetz)* of the Federal Republic contains a number of concepts which provide a framework for reform in West German society. With respect to the current local government reforms, the most important provisions of the Constitution are those found in Article 20 and Article 28.

Article 20, paragraph 1, declares the Federal Republic to be a "democratic and socially just *[sozialer]* federal state." This provision can be and frequently is interpreted to mean that the well-established—and by American standards very generous—network of social subsidies and social welfare programs must be available to all Germans in spite of the federal organization of West Germany. If necessary, these benefits in poorer states are to be financed in part by revenues from other wealthier states through the vehicle of the federal government or through a variety of fiscal transfers among the 11 states, among the counties within the states, and among the municipalities within each county. Thus Article 106, paragraph 2, calls for avoiding an overburdening of taxpayers in one part of the federation in proportion to other parts and for securing the unity of living conditions in the territory of the federation. Furthermore, Article 104a, paragraph 4, speaks of financial assistance being given to the states and their localities in order to maintain a general economic balance in the federation.

That part of the Constitution most directly related to local government is Article 28. Paragraph 1 requires that the constitutional order in the states—which includes the counties and municipalities—conform to the principles of a republican, democratic, and socially just political system under the rule of law *(Rechtsstaat)*. Paragraph 2 guarantees municipalities the right to regulate, on their own authority, all of the concerns of the local community within the framework of the laws. This means that, in contrast to Great Britain and most American states whose municipalities may act only according to national law or their charters, German municipalities enjoy a general power to act *(Allzuständigkeit)*. But this general power exists only in those areas in which the municipalities have legal autonomy, areas which in practice are becoming more narrowly defined as the areas of state and federal authority expand.[3] The counties, too, are guaranteed the right of self-government *(Selbstverwaltung)*.[4]

Increasingly German scholars and practitioners of public administration began to point to the discrepancy between constitutional theory and practice. How could social justice and equality in living conditions become a reality (in a rather perfectionist sense) when some states were considerably poorer than others and, in particular, when there were substantial differences in the administrative and service capacities *(Leistungsfähigkeit)* among the various counties and municipalities? The differences between the rural municipalities and the urban centers in the provision of a wide variety of public services—in spite of fiscal transfers and federal aid—were seen to have become especially intolerable in a modern, democratic welfare state. How could one speak of "genuine" local self-government, it was said, when most of the municipalities (in particular the smaller ones) and a large proportion of the counties were not in a position to meet their constitutional responsibilities, because of a lack of financial and administrative capacity to perform the tasks allotted them?

The constitutional argument for reform rests on the two principles of equality *and* freedom—or self-government—which are seen as complementary. Local governments in the Federal Republic, it is argued, must provide their citizens with a more uniform level of services than in the past. If they are not in a position to offer these services, then citizens will demand help from the state or federal governments, further eroding local self-government. It is assumed that such demands can be avoided only in a system of *reformed* local structures which are in a position to provide strong and effective, that is, "genuine" local self-government.

Relatively few scholars have seen the potential, if not real, contradiction which may have emerged between the constitutional principles of equality and social justice on the one hand and the guarantee of local self-government on the other. In the present-day context of the West German welfare state and changing technology, just what are "all of the concerns of the local community" which Article 28 gives the municipalities the authority to regulate? How does one separate these concerns from those of the state and federal governments? (Scheuner, 1973:6-7).

There has been considerable public discussion in West Germany in recent years of the concepts of equality and freedom and of numerous instances in which these are alleged to be in conflict. According to the political orientation of the particular politician, newspaper, journal, or commentator in the mass media, the concept of freedom has—or has not—been losing ground. Freedom versus "socialism" even became one of the issues in the October 1976 federal elections. Local government reforms have contributed to this discussion, as the debates in the state parliament of Lower Saxony, for example, demonstrate very clearly (Gunlicks, forthcoming). In these debates, spokesmen for the Christian Democratic party (CDU) accused the government controlled by the Social Democratic party (SPD) of carrying out municipal reforms in order to achieve increased equality at the price of local municipal freedoms and increased bureaucratization.

It is apparent that the German Constitution itself became a source of inspiration for those who advocated local government reforms. It should be noted, however, that the movement for local and regional reforms in Germany was also a reflection of general trends in Great Britain and much of Western Europe throughout the 1960s which placed, and continue to place, a strong emphasis on equality. When applied to regional and local governments, the concept of equality was viewed as requiring a reduction in the gap between services, benefits, and facilities provided in the urban as opposed to the rural areas and in one region as opposed to another. For that reason, among others, pressures for restructuring regional and local governments have been a European-wide phenomenon.

MODERN STRUCTURE AND PERFORMANCE

In addition to the above considerations, the vast literature on local government reform in Germany suggests that a modern, industrial welfare state requires a system of local government that is quite different from that which has come down from the past. The emphasis has changed from law and order and basic security to the provision of services (Niedersächsischer Minister des Innern, 1969:4). It has been argued that problems brought about by technology, such as automobile traffic, mass transit, or air and water pollution cannot be dealt with effectively or efficiently by local units whose boundaries do not permit the required coordination and focus of effort. New concentrations in business, commerce, and agriculture have long since expanded beyond the boundaries of municipal and county governments, yet these have an obvious impact on local governments (Rothe, 1968:2-3). Local and regional planning becomes increasingly difficult, it is argued, when overlapping and fragmented jurisdictions prevent the coordination of effort and unity of purpose needed for rational decision-making processes.[5] Modern local government, it is suggested, also requires highly qualified professional administrators, yet most villages and small towns have to depend on voluntary administrators because they are too small to justify hiring professionals. As a result, these small municipalities are viewed as turning more and more of their responsibilities over to the counties, which means a further weakening of "genuine" local self-government.

These problems—especially those of coordination and cooperation—can be observed in the relations between the medium-sized and large cities on the one hand and the surrounding villages and towns on the other *(Stadt-Umland-Problem)*. They are most serious and pressing in the most densely populated industrial regions of Germany *(Ballungs-, Verdichtungs-,* or *Verflechtungsgebieten)*. A number of attempts have been made to deal with these problems by creating a variety of special districts, intercommunity associations, joint authorities, and so forth, but it has been asserted that a lasting and effective solution can come only through major administrative, structural, and boundary reform. In other words, only a redistribution of responsibilities, changes in municipal forms, and a redrawing of municipal and county boundaries can deal adequately with the existing problems (Mattenklodt, 1972:9).

With few exceptions, advocates of reform argue that most local municipalities and counties are simply too small to perform the tasks which the Constitution (Weber, 1964:43-44; Rothe, 1968:17), modern conditions, and the people demand of them.[6] Patchwork solutions, such as joint authorities *(Zweckverbände)* and various associations of communities, are viewed as unsatisfactory short-term expedients which fail to come to grips with the real problems (Weber, 1964:20-21). They also are seen as violating the important principle of "unity of administration" *(Einheit der Verwaltung)*, according to which the administration of public affairs should be kept as simple and comprehensive—that is, with as little fragmentation—as possible, not only for the sake of efficient administration but also for the democratic purpose of pinpointing responsibility (Jochimsen et al., 1971:58-59).

There were some who argued against carrying these arguments too far. Thieme (1972b), for example, questioned the assumption of most reform advocates that "bigger" is automatically "better." He agreed that most villages lacked the administrative and service capacities required today and that they had an inadequate financial basis. Still, he suggested that villages of 1,000-5,000 possessed certain inherent values[7] and proposed that, instead of being consolidated or merged, they could be combined with others in an association of municipalities or communities for the purpose of dealing with regional problems (e.g., planning) or of providing more expensive services. As will be seen in the section below describing the actual structural changes made at the local level, five states did provide for associations of communities in certain instances.

The implementation of a more meaningful and responsible democracy at the local level became an argument which was used as a framework within which almost all of the other arguments were developed. It was suggested, for example, that measures to prevent the growing influence of state and federal governments over local governments and measures to promote public participation and democratic control of administrative agencies called for far-reaching local government reforms. Strong local units of government with strong financial resources—and that means larger population sizes— were viewed as necessary conditions for genuine local self-government. More citizens would be attracted to local institutions and be encouraged to participate in local public affairs if local governments acted on their own authority. The strengthening of local government under the principle of unity of administration, it was argued, would be a major contribution toward the strengthening of participatory democracy (Mattenklodt, 1972:70; Hessischer Minister des Innern, 1968:29).

These and other arguments made in favor of sweeping changes at the local level are, of course, hypotheses regarding certain cause-and-effect relationships between structure and performance. The arguments are not unlike those made by advocates of consolidation in the United States. They are perhaps less controversial, more plausible, and therefore more acceptable in Germany for a variety of reasons. For example, one could point to the population density of the country—60 million inhabitants in an area the size of the state of Oregon; the very different bureaucratic tradition; the more thoroughly developed and extensive welfare state tradition; racial homogeneity; the clearer distinction between cities and surrounding territories; and the susceptibility of Germans to arguments pressing for greater equality. The final section of this paper is devoted to an examination of these hypotheses with reference to the reforms which were implemented between 1968 and the present time.

STANDARDS

From the numerous arguments presented, there have emerged a series of standards which are seen as guidelines for the political and administrative officials who plan and implement structural and administrative reforms. In his exceptionally thorough, comprehensive, and influential book on administrative reform in Germany, Frido Wagener drew up a catalogue of 25 standards which had been discussed in the domestic literature on local government reform (Wagener, 1969:284-292). Eliminating contradictory or unrealistic standards (e.g., genuine reduction of administrative costs and personnel because of constantly rising demands for services, wages, and material improvements), he arranged the remaining 22 standards and 19 corollary statements into a general categorical scheme (pp. 312-313); see Table 1. This scheme, which is only partially presented here, became the basis for an elaborate process of quantification. More than 200 pages are devoted to a definition of terms and to calculations of optimum size for almost every conceivable public service or activity performed by state, county, and municipal governments (pp. 334-335, 463-466, 470-483).

On the basis of an individual review of 124 separate government responsibilities or tasks (pp. 328-462), from elementary schools to state planning agencies, Wagener presented his recommendations for minimum and optimum sizes for the administration of each service or activity (pp. 463-466). Then, dividing these 124 responsibilities according to governmental level, he calculated ranges and averages of population size for each task assigned (see note 3) to rural municipalities, medium-sized cities, large cities, counties, administrative districts (*Regierungs-* or *Verwaltungsbezirke,* for which there is no American counterpart), and states (pp. 469-483). Taking an overall average of all of the responsibilities for each level of government, he then arrived at an optimum figure or range for each unit of government. His conclusions are summarized in Table 2 (Wagener, 1969:504).

A review of the literature on local government reform in Germany

TABLE 1

STANDARDS FOR THE OPTIMUM ADMINISTRATIVE ORGANIZATION[a]

I. Effectiveness (Technical Standards)
  A. Economy
    1. Delimitation according to favorable[b] arrangement of population
    2. Reduction of administrative instances or levels
    3. Favorable[c] number of lower-level administrative units
    4. Financial, personnel, and structural balancing arrangements within the framework of one administrative unit
    5. Unity of administration
  B. Public Services Capacity
    1. Employment of full-time personnel
    2. High degree of specialization of personnel
    3. Good overview of the administrative unit

II. Integration Value (Political Standards)
  A. Strengthening Democracy
    1. Consideration of the feeling of community identity of population
    2. Channels of decision making regarding important administrative tasks
    3. Greatest possible decentralization of administrative tasks
    4. Simplicity and overview of administrative organization
    5. Congruence of administrative boundaries with those of other authorities—courts, electoral districts, organizations, planning agencies, etc.
  B. Securing Rule of Law *(Rechtsmässigkeit)*
    1. Binding nature of the law
    2. Legally regulated administrative behavior
    3. Judicial review of administrative measures

a. This scheme does not include the corollary statements mentioned in the text. Translation by the writer.
b. "Favorable" here appears to mean a population sufficient in size and distributed in such a way as to justify the activities of certain administrative agencies or the use of certain technical devices, such as computers (Wagener, 1969:317).
c. In this case "favorable" means, for example, 3 to 10 units in an association of municipalities or communities (Wagener, 1969:322).

and the results of the reforms in several of the territorial states —especially North Rhine-Westphalia—reveal the impact of Wagener's contribution. There were critical voices expressing some skepticism about his calculations or warning about a dogmatic dependence on them, but they were few in number and late in coming (Thieme, 1973; Mattenklodt, 1972:41). In some states the tempo of the reform measures was such that there was simply no time for the responsible parliamentary deputies to reflect on the principles underlying the reform goals (Krause, 1974:279).

Perhaps even more astonishing was how little attention was paid to the probable financial costs of the reforms (Laux, 1973:251-252) or to the question of where the money was to come from (Mattenklodt, 1972:207). On the other hand, a reduction of costs was never a

TABLE 2

FAVORABLE POPULATION SIZES FOR ADMINISTRATIVE UNITS
ACCORDING TO THEIR TASKS

Administrative Unit	Average Population Density (200-300 per square kilometer)	Below-Average Population Density (less than 200 per square kilometer)	Above-Average Population Density (more than 300 per square kilometer)
Rural municipalities	7,000 minimum	———————→	———————→
Medium-sized cities	40,000 minimum	———————→	———————→
Large cities (independent, i.e., "county-free")	200,000 minimum	———————→	———————→
Counties	170,000-380,000	130,000-280,000	210,000-480,000
States	8-14 million	6-10 million	10-7 million

primary goal of the reform movement, and in an absolute sense was rejected by Wagener (1969:295-296) in any case. Indeed, Werner Weber, the director of a state-sponsored study commission on local government reform, made it quite clear that there would be some significant increases in costs in the transition to the new system as the result of new buildings and facilities, new office equipment, specialized personnel, and a general upgrading of community services and facilities. He was convinced, however, that, because new sources of revenue would be tapped by the more professional and efficient administrative personnel, because decentralization would be made possible by strengthening municipalities, and because the number of special purpose authorities and their personnel would be reduced, one could expect that long-range economies would accrue from the reform. Weber formulated the problem of economy in the following way:

The attempt to avoid public expenditures if at all possible, an attitude found frequently in small municipalities, reflects a misunderstanding of the concept of frugality if important local responsibilities are then neglected. What is decisive is the principle of economy, according to which the greatest amount of modern administrative services is offered at the lowest possible technical and financial cost. [Niedersächsischer Minister des Innern, 1969:60; translation by the author]

LOCAL GOVERNMENT REFORM IN WEST GERMANY, 1969-1975

The 45th annual meeting of German legal scholars in 1964 (see note 5) is usually taken as the starting point for the current local government reform movement in West Germany. The Rhineland-Palatinate—the first state to conceive and carry out municipal, county, and district boundary and administrative reforms—had a head start when the minister-president (the German counterpart to the American governor) announced in May 1963 that his government intended to take up the question of local reform (Staatskanzlei des Landes Rheinland-Pfalz, 1966, vol. 1, p. 1). A comprehensive reform bill was first introduced early in 1965 and passed with major revisions in the summer of 1966. The 15th and last major bill for territorial reform was passed in the fall of 1973 (Staatskanzlei des Landes Rheinland-Pflaz, 1966, vol. 1, pp. 67 ff., 434 ff.; vol. 5, pp. 450 ff.). These measures included administrative reforms and changes in the statewide municipal, county, and district charters as well as boundary revisions and the consolidation or combination of numerous municipalities, courts, and other government bodies.

As a result of these reforms, the Rhineland-Palatinate reduced the number of its municipalities from 2,933 in 1965 to 2,366 in 1974; however, the number 2,366 requires elaboration, since it does not begin to reflect the actual change which took place. Twelve of the currently existing 2,366 municipalities, then as now, are independent, "county-free" cities with populations ranging from 37,000 to 200,000. Another 34 are "unitary municipalities" *(Einheitsgemeinden)*. These are combinations of villages functioning as a single unit, or a town surrounded by several consolidated villages. In function—but not in form—they are somewhat more comparable than the "county-free" cities to municipalities in the United States—excepting Virginia, which provides for city-county separation for municipalities above 5,000 in population (Gunlicks, 1973). Also, 2,320 municipalities, especially smaller towns and villages, have been formed into 166 associations of municipalities *(Verbandsgemeinden)*,[8] a kind of federation of villages designed to offer certain public services and facilities which the member communities would not be able to provide on their own. In contrast to the "unitary municipality," the member units retain their legal autonomy. Still, it could be argued that the Rhineland-Palatinate reduced the number of its municipalities from 2,366 to 212 (Buchsbaum, 1975:404) or, alternatively, that 166 units can be added to the remaining municipalities.

The number of counties was also reduced from 39 to 24, the average population increasing from 68,000 in 1964 (Rothe, 1968:76; Wagener, 1969:77) to 111,000 in 1974 (Buchsbaum, 1975:403-404). The five administrative districts *(Regierungsbezirke)* were reduced to three.

As a result of administrative reforms, the three new administrative districts were assigned 370 responsibilities previously assigned to various state ministries. At the same time, 316 duties were relinquished to the counties and independent cities. The 34 unitary municipalities and the 166 *Verbandsgemeinden* gained 237 new responsibilities for a total of 408 duties carried out on behalf of higher authorities or as a result of state and federal laws (Buchsbaum, 1975:404).

By 1974 the Rhineland-Palatinate had carried out a comprehensive territorial reform of its municipalities, counties, and administrative districts (Stich, 1969) and had implemented administrative reforms which brought about a considerable decentralization of responsibilities all the way from state departments to municipalities. As in other states, however, questions have been raised regarding the actual importance of the responsibilities which have been decentralized.

In three respects the Rhineland-Palatinate can be seen as the leader, as a standard, against which reforms in the seven other territorial states of the Federal Republic can be measured, because (1) it was the first state to plan and complete its reforms, (2) it was the first state to complete territorial reforms at all three levels —municipal, county, and administrative district (with the possible exception of one metropolitan area)—and (3) it was the first state to implement administrative reforms (decentralization). On the other hand, as the other states began to implement their territorial and administrative reforms, it became clear that the reform results in the Rhineland-Palatinate were rather modest by comparison (Buchsbaum, 1975, Table 6, p. 407; Wimmer, 1975, Tables 2, 4, and 5).

Table 3 shows that the consolidation of villages and towns in the Rhineland-Palatinate and Schleswig-Holstein was very modest indeed in comparison with the other territorial states.[9] Only Baden-Württemberg, the Rhineland-Palatinate, and Schleswig-Holstein failed to reduce the number of independent cities, although they did annex new territories to those already in existence. Baden-Württemberg, Bavaria, and North Rhine-Westphalia reduced the number of counties in their jurisdictions to a considerably greater extent than did the Rhineland-Palatinate; Lower Saxony, too, will probably reduce

TABLE 3

ORGANIZATION OF THE FEDERAL REPUBLIC OF GERMANY AFTER
10 YEARS OF TERRITORIAL AND ADMINISTRATIVE REFORM

State	Municipalities		Independent Cities		Counties		Administrative Districts	
	1965	1975	1965	1975	1965	1975	1965	1975
Baden-Württemberg	3,381	1,098[a]	9	9	63	35	4	4
Bavaria	7,097	4,178[b]	48	25	143	71	7	7
(Proposed changes)		(1,925)[c]						
Berlin (West)	1	1	—	—	—	—	—	—
Bremen	2	2	2	2	—	—	—	—
Hamburg	1	1	—	—	—	—	—	—
Hesse	2,693	591	9	6	39	24	3	2
(Effective Jan. 1977)		(417)				(20)		
Lower Saxonyt	4,245	1,017[d]	15	10	60	48	8	8
(Proposed changes)				(6)		(31)		(4)
North Rhine-Westphalia	2,362	370	38	23	57	31	6	5[h]
Rhineland-Palatinate	2,920	2,354[e]	12	12	39	24	5	3
Saarland	347	50	1	—	7	6[g]	—	—
Schleswig-Holstein	1,389	1,166[f]	4	4	17	11	—	—
	24,438	10,828	138	91	425	250	33	29
Probable final result		8,401		87		229		25

a. Reduces to 178 unitary municiaplities and 271 municipal associations (with 920 member municipalities).
b. Reduces to 4,123 unitary municipalities and 13 municipal associations (with 55 member municipalities).
c. Reduces to 768 unitary municipalities and 352 municipal associations (with 1,157 member municipalities).
d. Reduces to 271 unitary municipalities and 143 municipal associations (with 746 member municipalities).
e. Reduces to 34 unitary municipalities and 165 municipal associations (with 2,320 member municipalities).
f. Reduces to 100 unitary municipalities and 122 municipal associations (with 1,066 member municipalities).
g. Includes Saarbrücken metropolitan association.
h. This number will probably be reduced to 4.

substantially the number of its counties when its county reform program is completed, presumably in 1977. The administrative districts were least touched by the reform activity in the states.

Administrative reforms, in particular those which involved a decentralization through redistribution among local government institutions of a wide variety of responsibilities or tasks, have been completed or are in the process of being completed in all of the states except North Rhine-Westphalia and Lower Saxony (Buchsbaum, 1975:407). North Rhine-Westphalia began consideration of its administrative reforms in the fall of 1975; Lower Saxony, which had redistributed several hundred responsibilities in the process of carrying out its municipal reforms, will probably decentralize additional responsibilities upon completion of its county reforms.[10]

Tables 4 and 5 demonstrate that in terms of size the Rhineland-Palatinate also falls below average in its reforms of independent cities and counties. According to my research, it is apparent that one explanation for the smaller units in Bavaria, the Rhineland-Palatinate, and to a lesser extent Baden-Württemberg and Schleswig-Holstein, is political. All of these states are governed by the Christian Democratic Union (CDU) or, in Bavaria, the Christian Social Union (CSU—the CDU and CSU form a single parliamentary group in the national parliament, the *Bundestag*). The reform plans of the Christian Democrats have been generally less ambitious than those of the Social Democrats (SPD), which alone or in coalition with the Free Democrats (FDP) carried out the reforms in Hesse, Lower Saxony (see note 10), and North Rhine-Westphalia. Differences in administrative organization and tradition probably account as well for some of the dissimilarities among the various states in their reform efforts (Loschelder, 1969:225).

GENERAL IMPACT OF THE REFORM MOVEMENT: SOME TENTATIVE CONCLUSIONS AND HYPOTHESES

According to the arguments—or hypotheses—advanced in support of structural, boundary, and administrative reform at the local level in the Federal Republic, rather fundamental improvements in the quality, quantity, and efficiency of local services and administrations were expected to result from the implementation of a well-prepared and properly conceived reform plan in each of the eight territorial states of West Germany. Unfortunately, it is too soon to assess with any degree of confidence the relationship which exists between the reform arguments, as hypotheses, and their practical application. I am aware of only one study to date which focuses on the impact of the reforms in the terms of reform goals (Hirsch et al., 1974). One state, the Rhineland-Palatinate, which was the first to complete its reforms, commissioned in 1974 a comprehensive study of its reforms from the School of Public Administration in Speyer. (The Speyer report, available in the summer of 1976, appeared too late for analysis in this chapter.)

It is possible, nevertheless, to make some general comments about

TABLE 4
INDEPENDENT CITIES BY SIZE[a]

Population	Baden-Württemberg	Bavaria	Hesse	Lower Saxony	North Rhine-Westphalia	Rhineland-Palatinate	Schleswig-Holstein	Federal Republic
Less than 50,000	1	10	—	—	—	4	—	15(15)
50,000- 100,000	1	9	1(—)[b]	3(—)[c]	—	3	2	19(15)
100,000- 200,000	4	3	2(3)	5(4)	6	5	—	26(26)
200,000- 300,000	1	1	2(2)	1(1)	8	—	2	15(15)
300,000- 500,000	1	1	—	—	4	—	—	6 (6)
500,000-1,000,000	1	—	1(1)	1(1)	4	—	—	8 (8)
Over 1,000,000	—	1	—	—	1	—	—	2 (2)
Result as of July 1, 1975	9	25	6	10	23	12	4	91
Probable final result	9	25	(6)	(6)	23	12	4	(87)

a. The Saarland has no independent city.
b. After January 1, 1977.
c. Proposed changes subject to revision.

SOURCE: Buchsbaum, 1975:406.

TABLE 5
COUNTIES BY SIZE

Population	Baden-Württemberg	Bavaria	Hesse	Lower Saxony	North Rhine-Westphalia	Rhineland-Palatinate	Saarland	Schleswig-Holstein	Federal Republic
Under 100,000	2	45	3 (1)[a]	26 (—)[b]	–	11	1	–	88 (60)
100,000-200,000	19	26	13(10)	18(22)	4	13	1	9	103(104)
200,000-300,000	7	–	7 (7)	3 (7)	16	–	3	2	38 (42)
300,000-500,000	7	–	1 (2)	– (1)	10	–	1[c]	–	19 (21)
Over 500,000	–	–	–	1 (1)	1	–	–	–	2 (2)
Result as of									
July 1, 1977	35	71	24	48	31	24	6	11	250
Probable final result	35	71	(20)	(31)	31	24	6	11	(229)

a. After January 1, 1977.
b. Proposed changes subject to revision.
c. Saarbrücken metropolitan association.

SOURCE: Buchsbaum, 1975:406.

[187]

public reactions to the reforms and their apparent financial impact. On the basis of a survey of deputies conducted in one state parliament, I have also formulated a number of hypotheses which might guide further research efforts. These are presented in a section below.

CITIZEN PARTICIPATION AND ATTITUDES

One result of the local boundary reforms which was given very little attention in the 1960s and which contradicts, in part, the goal of democratization is the significant reduction in the number of municipal and county councillors. Before the first widespread boundary reforms were implemented in 1968, there were 276,000 elected councillors in the villages, towns, cities, and counties of the Federal Republic. Upon final completion of the reforms in 1978 there will be only about 146,000 local councillors, or 47% fewer elective offices available to encourage practical citizen participation in local governments (Wimmer, 1975).

There was, of course, no way to avoid a reduction in the number of councillors. To the credit of the state parliaments it must be noted that this aspect of the reform was not ignored. Indeed, one of the arguments in favor of the *Verbandsgemeinde* was that it offered a compromise between efficiency and capacity on the one hand and preservation of small municipalities and their councils on the other hand. Also in the constitutions (statewide charters) of the unitary municipalities *(Einheitsgemeinden)* and larger cities, provisions were made for subsidiary local or district councils *(Orts-* or *Bezirksräte);* however, some critics viewed these as little more than discussion groups dependent on the common council for all responsibilities and funds (Gunlicks, forthcoming).

Whether the consolidation of municipalities will lead in the long run to the goal of increased citizen interest and participation remains to be seen. By American and British standards, citizen participation in Germany was high even before the reforms, if one takes electoral participation as a measure.[11] If measured by the—in my opinion, problematic—civic norms used by Almond and Verba (1963, chap. 6) in their comparative study of political culture, citizen interest in local government is lower in Germany than in Britain and the United States. While there were some notable exceptions involving protest marches and referenda, the majority of citizens in the eight territorial states retained a somewhat skeptical and disinterested attitude toward the reforms (Conrady, 1973; Krabs, 1973).

LOCAL GOVERNMENT FINANCES

The local finance reform law of 1969 (which had little or no connection with the structural, boundary, and administrative changes discussed above) gave German local governments a share of the income tax (14%) and payroll tax. In 1974 these taxes supplied 50% of the revenue received by the municipalities. The tax on producing industries *(Gewerbesteuer)*, which before 1969 was the major source of local revenue, is now shared (i.e., about 40%) with the state and federal governments (Krumsiek, 1971:60-61). Real estate taxes, which the municipalities retained, amounted to about 16-17% of total revenues before 1969. In 1974 they accounted for 11.2% of all municipal revenues (Klein and Gleitze, 1975). The municipalities received 12.5% of the total West German tax revenues in 1974 but were responsible for 21.7% of all public expenditures (Klein, 1976:4), including about two-thirds of all public expenditures for capital improvements (Deutscher Bundestag, 1974). In spite of appeals from the local government associations, municipalities have not been granted the authority to add a local surcharge on payroll and income taxes (Krumsiek, 1971:61-62).

From 1962 to 1974 local government expenditures increased by 306%, while expenditures of the state and federal governments increased by 269% and 241% respectively (Klein and Gleitze, 1975:3). Rising personnel costs for well-paid local government employees have been responsible for most of the increases in local expenditures since 1969. Capital improvement expenditures have also been very substantial, and these have been affected strongly by high inflationary costs in the construction industry. By 1973 municipalities carried 46.3% (62.68 billiom DM) of the total public debt in Germany (Klein and Gleitze, 1975:4). This debt—and in particular the debts of the state and federal governments—increased sharply with the recession of 1974 and 1975 *(Der Landkreis,* 1976:9).

The current economic recession has led to generally stagnating revenues, while expenditures continue to increase.[12] For this reason and because some expenditures are tied to the administration of federal and state laws or set by contracts, the proportion of uncommitted revenues in the municipalities is estimated generally to range from 20% to as little as 6% of the total budget (Klein, 1976:6). As usual, the local government associations are predicting disaster for the future (Weinberger, 1976) and demanding a larger share of payroll and income taxes. The central government, which had to borrow a record 40 billiom DM due to the recession of 1975, has not been in any mood to meet these demands.

The above considerations suggest that, whatever local government reforms in West Germany have accomplished, they apparently have not brought municipalities and counties the degree of financial independence or support which these anticipated. A local finance reform—a redistribution of tax revenues among the various levels of government—was not one of the objects of the reform movement as discussed in this paper. Nevertheless, while warnings were sounded that, at least in the short run, the reforms might bring added costs, there were also numerous voices urging throughout the 1960s that local reforms would eventually bring considerable savings to local governments through rationalization, centralization, and economies of scale.

In the 1970s these arguments have become difficult to sustain. It has become increasingly apparent that there is a direct relationship between increased size of municipalities and higher personnel and service costs per citizen, although this relationship does not seem to hold for counties (Heemann, 1972:128-132). Studies pointing to such relationships contain ominous implications for those states which have insisted on creating larger-than-average municipalities. These implications were confirmed in at least one study, in which it was demonstrated that in North Rhine-Westphalia personnel costs had increased in reformed municipalities of 10,000 or more inhabitants at a higher rate than in unreformed municipalities (Hirsch et al., 1974). The obvious retort to this finding, that personnel costs increased because of improvements in the quality of services, cannot be demonstrated with the available evidence (Hirsch et al., 1974); however, these findings and the methodology of the study by Hirsch and associates have not gone unchallenged (Laux, 1973; Eichhorn, n.d.). Future studies will be required—and are now being undertaken—before one can declare with any confidence what the impact of local reforms has been on local government performance. What is clear at the present time is that these studies will not begin with the unchallenged assumption that reforms have reduced costs.

THE IMPACT OF MUNICIPAL REFORMS IN LOWER SAXONY:
SOME HYPOTHESES

In order to formulate some tentative hypotheses concerning the impact of municipal reforms in Lower Saxony (county reforms have not yet been implemented), the writer conducted a survey, by mail questionnaire, of that state's 155 parliamentary deputies. Most of the deputies who responded—about half—had been members of the state parliament during the 1970-1974 session when municipal reforms were pushed through by an SPD majority consisting of one vote. Most deputies are also county and/or municipal councillors while serving simultaneously in the state parliament (multiple officeholding is a rather common practice in Europe). It was my hope that the combination of political responsibility, high information level, and practical personal experience of these deputies would make them particularly well suited as a source of information about the impact of the reforms in Lower Saxony and, by implication, in the other states as well. At the same time I was aware of the probability of receiving responses which would be highly correlated with party affiliation.

The survey fulfilled both expectations. The SPD deputies were generally favorable in their assessments of the impact of the reform, since it was an SPD government which had pushed them through the parliament. The CDU deputies were generally very critical. In spite of this partisan factor, which, of course, affects the reliability of the responses, the results of the survey provide a foundation for a series of "hypotheses" concerning reform goals and reform results. There appears, for example, to be general agreement on the following propositions:

(1) Municipal reforms have led to improvements in the service and administrative capacities of municipalities.

(2) Municipal reforms have led to higher expenditures.

(3) Municipal reforms have led to increased bureaucratization, especially in smaller municipalities.

(4) Municipal reforms have increased the physical—and often the psychological—distance between citizens and their local governments; i.e., the goal of a *bürgernahe Verwaltung* has not been achieved.

(5) Public reaction to municipal reforms ranges from positive to negative but tends to be characterized by apathy, disinterest, and limited hostility.

Other propositions were less subject to agreement among the deputies. Thus the general assessment of the effects of municipal reforms on efficiency and economy of operations, on administrative simplicity, on reduction in the differences between services offered in the city and county, on equality among municipalities, and on city-suburban problems depends on the observer's selective perception as influenced by his ideological predisposition, party affiliation, and/or personal commitment to the reform plan.

To what extent the hypotheses will be confirmed or rejected by future studies remains to be seen. Some of the hypotheses need to be placed in context; for example, hypotheses (4) concerning the goal of a *bürgernahe Verwaltung* through administrative decentralization might yet be achieved once the *administrative* reforms—in contrast to the structural and boundary reforms—have been implemented. They were, after all, originally perceived as separate parts of a single package. Some of the negative feedback from the public regarding other reform goals may be based on misinformation or impatience among citizens. Alternatively, it might be the result of misconceptions about the probable effects of structural and boundary changes by those who espoused the reforms. In any case there appears to be little public enthusiasm for a reform movement which has occupied the time and effort of hundreds of scholars, leading civil servants and elected deputies in eight German states for 10 years or more.

CONCLUSION

It is possible that the apparent lack of public enthusiasm or interest in local government reforms is a reflection of what many in West Germany refer to as a general "reform fatigue" *(Reform-müdigkeit)*. Local reforms are merely one example of a whole series of reforms and rather abrupt changes in foreign *(Ostpolitik)* and domestic policies in Germany since the mid- and late 1960s. Disappointment with the gap between promise and performance and the passage of time have eroded much of the original enthusiasm and good will for reform which probably existed in the latter 1960s. Local government reforms have not been spared in the current sober and more skeptical view of social change.

It is also clear that local government reforms have been dealt a serious blow by the economic recession which began in 1974. Many of the expectations of the reformers were based on the assumption that the German economy would continue to grow and that public revenues would continue to increase above inflation, thus providing the financial base for additional qualified personnel and improved public services. At the moment at least, it appears that a major weakness in the conceptual framework of the local government reforms was the lack of attention paid to the question of providing adequate financing for the reform goals. Alternatively, financial problems may indicate that increasing costs are accruing as a consequence of the reforms. This is the thesis of the study by Hirsch and associates. However, according to initial newspaper reports on the study of the reforms in the Rhineland-Palatinate, not the reforms but other factors are responsible for rising expenditures (*Frankfurter Allgemeine Zeitung,* April 27, 1976, p. 4).

By the middle of 1976 it was apparent that most of the technical goals of the local reform movement, such as a substantial increase in the size of municipalities and counties, had been met; others, such as administrative decentralization, were only partially and grudgingly implemented by higher administrative agencies or were still being considered. Yet the administrative reforms were crucial to the original concept. Villages which were assumed to be too small to provide sufficient services and facilities were to be annexed or joined together in a variety of structural forms. Through administrative decentralization citizens would be able to go to the new local administration for routine administrative matters rather than having to travel to the county seat. For the more complicated questions they would be able to turn to the new county administration rather than having to go through state agencies. To the extent that administrative reform—i.e., administrative decentralization—has not been carried out, the reform programs are incomplete. Whether the original promises in the plans of reform advocates will be realized once all aspects of the reforms have been completed in the eight territorial states remains to be seen. It would be surprising if there were not some disappointments.

NOTES

1. West Germany is a federation of 10 states (*Länder,* plural; *Land,* singular), including the city-states of Bremen and Hamburg. The 11th state, the city-state of West Berlin, is not legally a part of the West German Federal Republic; however, for most practical purposes it is an integral part of West Germany.

2. Actually, it can be said that the number of municipalities will be around

3,330—including 87 independent large cities—by the end of 1977, if one takes into account the fact that about two-thirds of the present 11,000 reformed municipalities have been or will be brought into various cooperative combinations and associations of municipalities (see Table 3).

3. In the Federal Republic the tasks and responsibilities of the local governments are divided into two general categories. In the municipalities, for example, the first category consists of those activities and affairs for which the municipalities are themselves responsible, but not in every case entirely autonomous, and which directly concern the local community. This *eigener Wirkungskreis*, this "own sphere of activity," may include those activities and responsibilities which the municipalities have assumed voluntarily and over which they exercise autonomous authority, such as theaters, parks, and playgrounds; it may include the activities which the municipalities have voluntarily initiated and over which they do not exercise autonomous authority, but which are regulated by state or federal laws, such as local public savings banks *(Sparkassen)*; and it *must* include certain activities and responsibilities made *obligatory* by state law *(Pflichtaufgaben)*, such as schools, public welfare services, fire protection (but not police protection, which is a state responsibility), and public transportation facilities. The second category *(übertragener Wirkungskreis)* consists of those activities and affairs administered by the local governments on behalf of the state and federal governments. Thus the local civil servant administers not only the ordinances passed by the local council, but also many of the laws passed by the state and federal parliaments.

4. The question of whether Article 28 and the guarantee of local self-government to the municipalities and counties forbids the elimination of individual municipalities and counties has been settled by German courts in favor of the states and their right to annex or consolidate. The courts have ruled that municipalities and counties are constitutionally protected as *institutions*, not as individual units within their traditional boundaries; however, some recent decisions have upset a number of annexations which in the judgment of the courts denied due process of law or were not sufficiently justified by the circumstances (Thieme, 1972a; for recent decisions, see *Das Parlament*, 1976, and *Der Städtetag*, 1976).

5. Some scholars date the beginning of the current local government reform movement from the annual meeting *(Juristentag)* of legal scholars in 1964, when the theme for discussion was problems of local government structures and planning (see Weber, 1964). The *Juristentag* is generally composed of legally trained experts (mostly professors and high-level civil servants) in public law and public administration.

6. In 1965 about 70% of the 24,500 municipalities in West Germany had fewer than 1,000 inhabitants (*Statistisches Bundesamt*, 1966:36).

7. In the state parliaments in which the reforms were debated and passed, the question of size became a partisan issue, with the Social Democrats (SPD) generally advocating larger units, the Christian Democrats (CDU) smaller ones (for Lower Saxony, see Gunlicks, forthcoming).

8. The states of Hesse, North Rhine-Westphalia, and the Saarland have provided exclusively for the consolidated, unitary municipality *(Einheitsgemeinde)*, which normally has a population of 7,000 or more. In contrast, the Rhineland-Palatinate, Baden-Württemberg, Bavaria, Lower Saxony, and Schleswig-Holstein have all attempted through the institution of the *Verwaltungsgemeinschaft, Samtgemeinde,* or *Amt* to find an arrangement by which villages and towns could perform certain services cooperatively without giving up their autonomy entirely. These associations of municipalities are generally an alternative to the unitary municipality only in rural areas (see Buchsbaum, 1975:405).

9. It should be remembered, however, that in these two states in particular attention was focused on the formation of associations of villages and towns rather than on the creation of unitary municipalities through consolidation.

10. County reforms in Lower Saxony were dropped once in 1970 due to the collapse of

the "Grand Coalition" between the SPD and CDU. The reform package introduced in the fall of 1975 and programmed for passage through the state parliament in April 1976 will be delayed and probably revised considerably as the result of the collapse in January and February 1976 of the SPD-FDP coalition and the formation of a minority CDU government in Hannover. Partisan politics have been an important factor in the local government reforms of Lower Saxony (Gunlicks, forthcoming).

11. Electoral participation in German local elections ranges from 70% to 80% as opposed to about 40% in Great Britain (Pridham, 1973:456). It is generally even less in the United States.

12. The increasing debt of the municipalities, brought about primarily by the economic recession, has prompted a continuing debate in West Germany concerning the desirability and feasibility of having certain public services performed by private contractors. In spite of the caution which exists in conducting such experiments in general, numerous cities —especially those with fewer than 10,000 inhabitants—have turned over their garbage collection to private concerns (*Frankfurter Allgemeine Zeitung*, March 27, 1976, p. 2).

REFERENCES

ALMOND, G., and VERBA, S. (1963). The civic culture: Political attitudes and democracy in five nations. Princeton, N.J.: Princeton University Press.

BUCHSBAUM, R. (1975). "Gliederung des Bundesgebietes nach 10 Jahren Gebiets- und Verwaltungsreform." Der Landkreis, 10:399-408.

CONRADY, H.W. (1973). "Das Interesse des Bürgers an der kommunalen Gebiets- und Verwaltungsreform." Der Landkreis, 3:76-77.

Deutscher Bundestag (1974). Drucksache 7/2409.

EICHHORN, P. (n.d.). "Zielsetzungen und Erfolgskriterien der Verwaltungsreform." Unpublished manuscript, Hochschule für Verwaltungswissenschaften, Speyer.

GUNLICKS, A. (1973). "City-county separation and local government reform in Virginia." Virginia Social Science Journal, 8(November):47-61.

——— (forthcoming). "Parteipolitik und Gemeindereform in Niedersachesen."

HEEMANN, H. (1972). Der Einfluss der Grösse eines Verwaltungsbezirkes auf die Personalkosten der Verwaltung (Schriften zur Verwaltungslehre, Heft 12). Bonn: Carl Heymanns Verlag.

Hessischer Minister des Innern (1968). Verwaltungsreform in Hessen. Wiesbaden.

HIRSCH, H., STICKER, J., and SCHUNK, D. (1974). "Die Personalkosten der Gemeinden in Nordrhein-Westfalen 1967-1971 unter dem Einfluss des ersten Neugliederungs-program." Städte- und Gemeinderat, 6:202-212.

JOCHIMSEN, R., KNOBLOCH, P., and TREUNER, P. (1971). Gebietsreform und regionale Strukturpolitik: Das Beispiel Schleswig-Holstein. Opladen: Leske Verlag.

KLEIN, R. (1976). "Gemeindefinanzbericht 1976." Der Städtetag, 1:2-14.

KLEIN, R., and GLEITZE, J. (1975). "Gemeindefinanzbericht 1975." Der Städtetag, 1:2-15.

KRABS, O. (1973). "Gebietsreform als Informationsproblem." Der Landkreis, 4:131-132.

KRAUSE, P. (1974). "Die Gebiets- und Verwaltungsreform im Saarland." Archiv für Kommunalwissenschaften, 13(2):277-290.

KRUMSIEK, R. (1971). "Zwischenbilanz zur Gemeindefinanzreform." Archiv für Kommunalwissenschaften, 10(1):54-69.

Landkreis, Der (1976). "Zur Finanzsituation der kommunalen Körperschaften." 1:9-11.

LAUX, E. (1973). "Die kommunale Gebietsreform: Ein Literaturbericht." Archiv für Kommunalwissenschaften, 12(2):231-256.

LOSCHELDER, W. (1969). "Verwaltungsreform—Eine Bilanz über erste Ergebnisse." Die Offentliche Verwaltung, 22(7):225-230.

MATTENKLODT, H.F. (1972). Gebiets- und Verwaltungsreform in der Bundesrepublik Deutschland. Münster: Verlag Aschendorff.

Niedersächsischer Minister des Innern (1969). Verwaltungs- und Gebietsreform in Niedersächsen: Gutachten der Sachverständigenkommission für die Verwaltungs- und Gebietsreform (Weber Kommission). Hannover.

Parlament, Das (1976). February 7, no. 6, p. 9.

PRIDHAM, G. (1973). "A 'nationalization' process? Federal politics and state elections in West Germany." Government and Opposition, 8(October):455-472.

ROTHE, K.H. (1968). Der ideale Verwaltungsbehördenaufbau in den Bundesländern. Göttingen: Verlag Otto Schwartz.

SCHEUNER, U. (1973). "Zur Neubestimmung der kommunalen Selbstverwaltung." Archiv für Kommunalwissenschaften, 12(1):1-44.

Staatskanzlei des Landes Rheinland-Pfalz (1966-1973). Verwaltungsvereinfachung in Rheinland-Pfalz: Eine Dokumentation (5 Bände). Mainz: Deutscher Gemeindeverlag.

Städtetag, Der (1976). "Urteile des Verfassungsgerichtshofs NW zur Neugliederung." Der Städtetag, 1:31-33.

Statistisches Bundesamt (1966). Statistisches Jahrbuch für die Bundesrepublik Deutschland. Stuttgart: Kohlhammer Verlag.

STICH, R. (1969). "Rheinland-Pfalz—Beispiel einer durchgreifenden Verwaltungsreform." Die Offentliche Verwaltung, (7):236-239.

THIEME, W. (1972a). "Bundesverfassung und Gemeinderecht." Juristenzeitung, 15/16 (18 August):478-482.

——— (1972b). "Vom Nutzen kleinerer Gemeinden." Archiv für Kommunalwissenschaften, 11(2):358-364.

——— (1973). "Die magische Zahl 200,000." Die Offentliche Verwaltung, 26(13):442-446.

WAGENER, F. (1969). Neubau der Verwaltung. Berlin: Duncker und Humblot.

WEBER, W. (1964). "Entspricht die gegenwärtige kommunale Struktur den Anforderungen der Raumordnung?" Verhandlungen des 45. Deutschen Juristentages (Band 1). Munich and Berlin: C.H. Beck'sche Verlagsbuchhandlung.

WEINBERGER, B. (1976). "Ein markanter Einschnitt." Der Städtetag, 1:1.

WIMMER, S. (1975). "Nur rund ein Drittel der Gemeinden bleibt." Der Städtetag, 9:463-467.

7

Police Reforms in The Netherlands: Expectations About Structure and Performance

RUUD J. VADER

☐ IN ENHANCING UNDERSTANDING of societal phenomena, one looks for aspects of a case that are comparable with those of other cases and explainable in broad terms. It is in this sense that the current Police Reform proposals in the Netherlands are important. There is nothing unique about the problems that the Dutch face in coping with developments common to most industrialized welfare states—including rising crime rates, the increasing mobility of criminals, the ungovernability of large-scale public service organizations, and the growing discrepancy between the demands made upon those organizations and their capacity to respond to such demands.

AUTHOR'S NOTE: *This article would have been impossible without the support, criticism, and materials that I received from many. I am especially indebted to P. Van Dijke, Secretary General of the Dutch Ministry for Home Affairs; H.V. Holtus of the Ministry of Justice; J.M. Polak, Governmental Commissioner for revision of the Police Act; H.R.G. Veldkamp of the Association of Dutch Municipalities; F.P. Bish and V. Ostrom of the Workshop in Political Theory and Policy Analysis at Indiana University; D. Goetze of the Political Science Department at Indiana University; and H.P. Vonhogen of the Erasmus University at Rotterdam, The Netherlands. Their ideas and comments have been helpful. As usual the responsibility for this article remains my own.*

Commonalities exist not only in the problems that the Dutch and other industrialized nations face but in their diagnosis and in the basic solutions advanced to solve these problems. Take, for instance, the notion that reorganization (and especially centralization and consolidation) is essential to the solution of many of the current problems. As is the case in many other countries, the Dutch are currently considering proposals to restructure the organizational arrangements used for the delivery of essential public services. Unlike many other countries, however, these reorganization proposals would affect the entire structure of public administration in the Netherlands.

The Police Reform proposals represent an important, albeit somewhat neglected, aspect of the reorganization currently under consideration. These proposals (and others affecting public administration in the Netherlands) stand a good chance of being adopted —adoption requiring no more than a simple majority vote of the Parliament. This article first describes the Police Reform proposals, their historical and contemporary context, and the expected benefits arising from them. Next, some possible effects of the reorganization are analyzed, and, finally, some suggestions for further research are made.

THE CURRENT STRUCTURE

A fundamental principle underlying the governing of the Netherlands has always been that no administrative task should be placed on a level higher than that at which it can be adequately performed. This principle is manifest in the tendency to decentralize governmental functions and power whenever possible. Consequently the municipalities have traditionally been, and still are, the basic units of local government.

The provision of police services used to be a strictly municipal affair. Until the mid-19th century, municipalities were the only producers of police services in the Netherlands. However, changing demands, like those induced by urbanization, the increased mobility of the population, large-scale strikes, and other manifestations of social unrest, created problems that could not be dealt with by municipal police alone. With industrialization and population mobility, many smaller, more rural communities also began to face an increasing need for law enforcement services on a routine basis.

Hence, an administratively decentralized National Police Corps was established in the mid-19th century to provide general police services to small towns and rural areas without their own municipal police and to deal with organized crime and other large-scale threats to public safety in urban areas and on a nationwide basis.

The duties of the municipal and National Police are described in the Police Act of 1957. These are

> to act in subordination to the competent authority and, in accordance with the prevailing rules of law, to maintain law and order and give assistance to those who need it. [Article 28]

There are two "competent authorities" within the executive branch of the Dutch government. One is the Department of Justice with the Attorneys General as regional representatives. The other is the Ministry for Home Affairs with Queen's Commissioners as the provincial representatives and with mayors as the local representatives. The way in which their respective jurisdictions is delimited is complex and has been the source of much controversy. This controversy relates to the two dualisms that characterize the organizational structure of the Dutch police. These are:

(a) A dualism in "rule-making authority" which is manifest in the two kinds of police that exist: National and municipal.

(b) A dualism in "operational authority" related to the two basic functions performed by the police: maintaining public security and combating crime.

RULE-MAKING AUTHORITY

Rule-making authority pertains to the power to nominate, appoint, or promote officers, to allocate financial and material resources, and to issue regulations determining internal organization and communications. Part of this power rests (for both the National and municipal police) with the central government. The Minister for Home Affairs and the Minister of Justice are authorized to make decisions—separately or in cooperation—concerning such matters as legal status, oaths of office, ranks, pay, discipline, qualifications required for appointment and promotion, recruitment, training, uniforms, arms, medical examination. They also have the right to nominate candidates for the highest ranks in the force. In Figure 1, this "core" authority is represented by dotted lines.

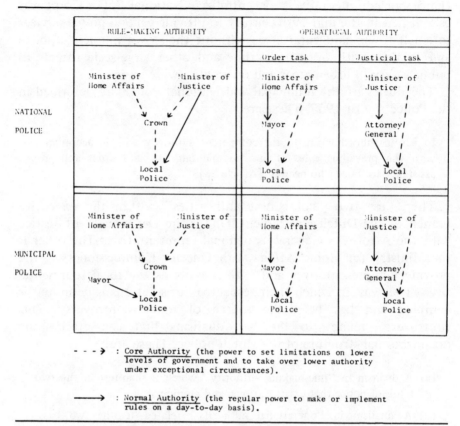

Figure 1: CURRENT AUTHORITY STRUCTURES FOR NATIONAL AND MUNICIPAL POLICE FORCES

Within the limits established by these central directives, rule-making authority for the National Police lies with the Minister of Justice and for the municipal police with city mayors. (The size of the city is generally the criterion for deciding whether or not a city will have its own police force.) Comments made by Mayor Schols of Venray after his city had acquired a municipal police force, demonstrate the immediate importance that is attached to this rule-making authority:

Mayor Schols sees as the most important improvement the fact that the police are now hierarchically subordinate to him and not to someone outside the municipal apparatus. It is possible he says, to have good relations with a group or district commander of the National police, but if he does not go along with your policy, you're through. *You depend on him.* You have to advance very strong arguments in order to achieve, for instance, that a specific person will or will not be hired. Now the municipal administration (Mayor + City Council, RJV) *can direct the police as it desires.* [Schols, 1974; italics mine]

It must be noted that the emphasis here is on one part of the dual police function, that for which the mayor is directly responsible.

OPERATIONAL AUTHORITY

Police services fall in either of two categories. One is that of maintenance of order and public security. The other concerns law enforcement and sanctioning tasks. For each category the production unit is responsible to a different competent authority, independent of where the rule-making authority for that unit is located. The corresponding divergence in chains of command is reflected in the second and third column of Figure 1.

The task of enforcing the law—especially the investigation of crimes and violation of laws, including apprehension of criminals—is in the Netherlands referred to with the adjective *justiteel.* This term, which is translated here as *justicial,* reflects the competent authority to whom the police are subordinate in the performance of this task—the Minister of Justice and those to whom he has delegated power. In performing justicial tasks, police are under operational command of this branch of the executive. This is the case for both the National Police and the municipal police. Indirectly, justicial police services contribute to order maintenance in a broad sense. They concern violations of rules, and they enable the prosecution to function and have suspects brought before representatives of the judiciary.

The *order task,* on the other hand, entails activities like regulation of traffic, neighborhood patrol, and returning juvenile runaways to their homes, to name just a few. Performance of the order task has a direct impact on public security and is executed under the operational authority of representatives of local, provincial, and national governments. Again this applies to National and municipal police alike.

As can be seen in Figure 1, the chains of operational command are the same for both the National and municipal police. In both instances, mayors are responsible for the maintenance of order and the attorneys general, as regional representatives of the Minister of Justice, for the justicial task.

Although the structures of operational authority are similar for the order and justicial tasks, their substance differs, both with respect to democratic control (discussed below) and with respect to the degree and type of direct involvement by authorities in task performance. The justicial police task is not only the attorney general's responsibility, but it is his professional specialty as acting director of units of the National Police. Mayors, on the other hand, are generally not specialists in the order task. Nor does order mainenance constitute the major part of the mayors' workload as does the justicial task for the attorneys general.

DEMOCRATIC CONTROL

The Ministers of Justice and of Home Affairs are politically responsible for the police to the parliament. There is, however, no direct democratic control over the police on the local level. The competent authorities at that level, the mayors and attorneys general, are not elected but appointed by the Crown. They are answerable only to higher authorities, and mayors are nearly autonomous with respect to the local task of maintaining public order. Since 1969 mayors have become accountable for this task to their city councils. The council, however, cannot apply any direct sanctions. It cannot veto or overrule a mayor's decision. It can only try to persuade him or appeal his decision to provincial or national authorities. Although the mayor is obliged to provide the council with information on any aspect of the police tasks it requests, he can withhold this information if he deems its dissemination contrary to the public interest. Thus, even under conditions where there is a cooperative relationship between the mayor and the city council, one can, at the most, speak of a democratic *influence* in the control of police at the local level.

CHANGING DEMANDS ON OLD INSTITUTIONS

The discussions on police constitute only a part—although an important part, in the eyes of many—of the concerns of the Dutch authorities charged with the complex and difficult task of conceptualizing new institutional arrangements. The demands that these designs have to meet are manyfold; sometimes they are contradictory. Ideally it is hoped that they will provide solutions to many, if not all, of the practical problems that modern society poses for the production of public services. It is also hoped that they will allow for more direct popular participation, while retaining—or even increasing—uniformity in the provision of many public services. These prescriptions derive from the principles of "democratization" and "equitable law enforcement" respectively.

To understand the nature of many of the practical problems involved, one must bear in mind that the consumers of public services have, in the last decades, become less identifiable in terms of municipal or even regional boundaries. Many decisions on a local level (i.e., those concerning housing, public transportation, the establishment of new industries, hospital and ambulance services, the quality of the environment and public security) have come to have an impact that reaches beyond municipal boundaries. The responsibility for protecting these wider interests is not readily transferred to higher levels of administration. Although some services are traditionally provided, or at least coordinated, at the provincial level (such as utilities and regional zoning), the provinces, being predominantly institutions of supervision and appeal, lack the necessary administrative capabilities for meeting the new demands. In addition, they are considered to be too large and based on obsolete geographical conditions.

During the search for a new and viable institutional design, temporary solutions have been found in intermunicipal cooperation. Most of these arrangements are ad hoc but sometimes have been formalized in more or less permanent intermunicipal agreements. The latter include some large-scale experiments with regional councils. A well-known example exists in the "Rijnmond" region, the industrial and harbor area surrounding the city of Rotterdam. Notwithstanding the many merits that these arrangements have over less institutionalized forms of cooperation, their basic weakness lies in their confederative nature. Those representing the intermunicipal interests have only delegated authority, and those having primary authority (i.e., the cities) often regard their narrower local interests as more important than regional interests.

The changing basis for identifying the consumers of many public services has not left police organizations unaffected. Increases in commuter and other types of traffic, in crime rates, in drug abuse, in hard crimes like hijacking, kidnapping, and armed robbery, in pollution and concentration of certain categories of the population in urban and suburban areas have intensified the demand for more specialized and more differentiated police services.[1]

An important qualitative change has also occurred with respect to the underlying principles of the "order" which police are supposed to maintain. The traditional concept of "legal order" has been supplemented by notions derived from social and humanistic values. Instead of viewing a citizen merely as a subject of legal rules, his or her personal circumstances are increasingly taken into account. For instance, the treatment of convicts has moved from the realm of pure punishment to individualized "rehabilitation" in which parole, probation, reeducation, and guidance are the catchwords.

Generally there has been a transition toward more consumer-oriented provision of public services.[2] This has been a painful and controversial process of which student movements have been rather radical advocates. These developments have left the police in a situation where they have to maintain an order about which there is little consensus. The question facing institutional "architects" in the Netherlands is, then, one of how police can be organized in such a way that they are capable of responding to these changes.

At an early point Dutch officials concluded that a partial restructuring of the governmental machine would be of little avail; it was felt that a basic overhaul was in order. The current plans for administrative reform, which stands a good chance of being adopted, call for elimination of the existing eleven provinces, replacing them with 26 (new style) provinces. These provinces are to assume many of the tasks that have become too burdensome to be effectively and efficiently performed by the cities. These tasks include, among others, regulation of traffic, public transportation, health care, large industries, large-scale recreational projects, culture, protection of nature and landscape, police, fire fighting, pollution control, and regional zoning (Algemeen Politieblad, 1975).

The most recent plans for police reforms are based on the new provincial system. These plans are assumed to satisfy at least three factors that have been of central concern in the discussions on police reforms. First, the regional governing apparatus should have sufficient power to perform its task. Specifically, control over the production of police services on a regional level should be adequate and not dependent on cooperation between more or less autonomous local governments. Second, the regional governments should be susceptible to democratic control at a regional level. The current provincial council, being predominantly a ratifying and appeal institution, needs fundamental restructuring. Finally the plans should allow for the establishment of police agencies with a minimal strength of 400 to 500. It is generally felt that this strength is the minimum necessary for the sophisticated, effective, and efficient operation of police units (Van Agt and Geertsema, 1972:6, 9; Vereniging Nederlandse Gemeenten, 1972:25; Van Agt and DeGaay Fortman, 1975:6).

THE POLICE REFORM PROPOSALS

There are a number of other factors that cause great concern (Van Agt and Geertsema, 1972:5-7, Procureurs-Generaal, 1974:1-3). One is the way in which mass communications transforms local crimes and disturbances into matters of regional and national significance. This has generally made these events more visible to the public, thereby increasing the pressure on police to do something about crime.

Another practical problem is represented by the crime statistics. The percentage of reported crimes solved by the police has decreased from 40.6% in 1970 to 31.8% in 1974. The reported crimes show a yearly increase of roughly 10%. Even if one takes into account the facts that the rate of increase in crime rates is declining (13.1% for 1971-1972 and 9.6% for 1973-1974) and that statistics may not precisely represent the actual situation, one does not need much imagination to see why there is criticism of police performance.

A common reply of spokesmen for police organizations is that these agencies are understaffed and often have insufficiently trained personnel. (The average size of a local detachment of the National Police in a rural area is approximately 6; the average size of a municipal force is approximately 115.) Related to the problem of size and understaffing is the lack of sophisticated equipment. Discussion of these problems is occurring in a context where the status of police officers has decreased considerably as a result of changed value orientations. This has made the task of recruiting prospective officers even more difficult. Moreover, the change in emphasis away from a purely legal and toward a social "order" has had an effect on the willingness of young people to apply for positions with the police where this change has not yet been internalized.

A more direct problem is the controversy over the predominance of the order task over the justicial task. The shift in emphasis from sanctions to rehabilitation of convicts and changed notions of "order" have led many mayors to concentrate on the former at the expense of the latter. This is, of course, troublesome for the authorities responsible for the justicial tasks. In response, they are lobbying for a well-defined influence in rule-making for all police agencies (Procureurs-Generaal, 1975:1-6; 1974:3-10).

SOLVING THE PRACTICAL PROBLEMS

The key solution proposed for these practical problems is consolidation. There is, given the Dutch tradition to the contrary, a surprising absence of disagreement about this solution. Consolidation would lead to the establishment of 26 provincial police corps with a minimal strength of approximately 900.[3] The minimal strength would be sufficient to provide adequate recruiting services, training, and task specialization. The larger areas to be served by the new corps would establish a sufficient base for purchasing and operating sophisticated equipment. Such an enlargement of scale, then, would satisfy "The demands which . . . must be made of the police from the point of view of organization, training, composition and management" (Van Agt and Geertsema, 1972:6).

UPHOLDING THE PRINCIPLES

In contrast to the consensus evident in the diagnostic assessments of and the solution posed for the practical problems, democratization and equitable law enforcement are the subject of much debate. The reason for the debate is the unfortunate situation that advancement of one principle has a direct negative impact on the other. Equitability is assumed to require uniformity—and hence centralization—in the production of police services.

What does democratization mean in this context? The following quotation indicates its content and its limitations:

> The idea of bringing the administration and the citizenry as close together as possible leads to the conclusion that it is worthwhile to strive for protection of interests, of clear and immediate importance to the individual, [at a level of administration which is] as near to him as possible, and under democratic oversight. [Van Agt and DeGaay Fortman, 1975:2]

The Dutch word *controle* is deliberately translated as "oversight" so as not to give it the connotation of power which is present in the

English word "control." What the ministers mean by democratic oversight is identical to what has been described above as democratic *influence.* "Only when reasons of effectiveness or the nature of the interests concerned—for example, national defense—make it necessary, should the protection of interests take place on a higher level" (Van Agt and DeGaay Fortman, 1975:2).

The assumption of the ministers is that democratization in this sense requires decentralization of governmental function. However, in addition to the two intrinsic constraints mentioned—effectiveness and the nature of the interests—there is another constraint that imposes limits on the extent of decentralization. That is the principle of equitable administration of justice. This principle is exemplified in the unity underlying both the centralized organization of the judiciary and in the legal procedures prescribing how justice must be administered. Maintaining a dynamic balance between this unity and the situational variations that bear on maintaining the public order in a broad sense is the task of the attorneys general and, on another level, the mayors.

THE STRUCTURE

Before looking at the degree to which the practical and principal considerations correspond with the proposed design, a description of the latter is in order. The emphasis is on the provision of "regular" police services for local communities. The specialized and mostly centralized police forces, such as river and harbor police, the central information service, and the crime laboratories, are mentioned only in passing. No attempt at completeness is intended.

The proposed restructuring of rule-making and operational authority is depicted in Figure 2. In comparison with the existing structure there are three fundamental changes:

1. There will be one type of police (i.e., provincial) instead of two (National and municipal). (This will eliminate one dualism: that of rule-making authority.)

2. Roughly 25% of the police forces will be decentralized. The Minister of Justice will lose his direct rule-making authority over this part of the police forces.

3. The urban police forces, roughly 75% of the police forces, will be centralized into provincial corps. Here it is the mayor who will lose his direct rule-making authority.

Although the structure of operational authority remains prac-

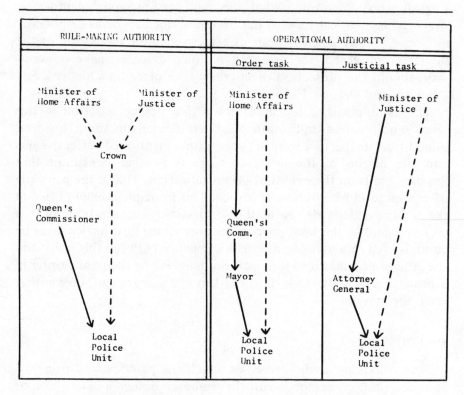

- - - →   : <u>Core Authority</u> (the power to set limitations on lower
levels of government and to take over lower authority
under exceptional circumstances).

———————→   : <u>Normal Authority</u> (the regular power to make or implement
rules on a day-to-day basis).

**Figure 2: PROPOSED AUTHORITY STRUCTURES FOR PROVINCIAL POLICE FORCES
AT THE LEVEL OF LOCAL UNITS**

tically unaltered, the Queen's Commissioner will be more directly
involved in the order task. He actually obtains direct operational
authority in relation to the provincial aspects of the order task; this
sometimes he does not have under prevailing conditions. The
ministers also recommended that the desirability of extending the
"special" power (associated with important crimes, large-scale
disturbances, emergencies, etc.) of provincial and national authorities
be evaluated. By proposing to put two commanding authorities on a
relatively equal footing—neither the attorney general nor the mayor
will have rule-making authority—two old problems are retained. The

first is reconciliation of the different demands placed on police by different operational authorities; the other concerns the synchronization of rule-making with operational authority.

Apart from a change in what may be called derivative constitutional authority,[4] these problems of synchronization have traditionally been inherent in the structure of the National Police organization. The (not always satisfactory) solution has been: consultation and cooperation between the parties concerned on a voluntary basis. In response to both the increase in scale and conceptual intensification of these problems,[5] the ministers have proposed an institutional arrangement of so-called "permanent triangular consultation" on the provincial level. The three "corners" of the triangle are (1) the rule-making authority (the provincial Queen's Commissioner), (2) the operational authorities, and (3) the chief of police. A similar triangle is suggested at the national level of government.

The basic rules determining relationships between parties in the triangle, are:

- Each party has the right to advise the other parties.
- Each has a right to be consulted.
- Each has the right to demand information about specific facts and situations.

Implied, also, are the corollary obligations to substantiate these rights.

How will these changes affect the provision of police services in urban communities? It is evident that powerful provincial governments will be established at the expense of mayoral powers. The mayor's position vis-à-vis the police will be considerably weaker. He will have no rule-making authority, and he will have a reduced operational authority. The position of local chiefs of police will be weakened in that they will become subordinate to the provincial chiefs of police. This of course decreases the mayor's effective operational power even further. Moreover, each mayor will have to compete for police services with some 30-odd other mayors. "Permanent triangular consultation" is the proposed means of avoiding and solving potential problems. It is a functional substitute for a more decentralized production of police services.

A PARTIAL ANALYSIS OF THE PROPOSALS

This analysis is partial in that it concentrates on aspects of the proposals that deal with the provision of "regular" police services at the local and provincial levels of government. It focuses on the degree to which proposed changes can be expected to bring about desired effects.

The diagnosis underlying the Police Reform proposals rests on two findings:

(1) Municipal police forces, under conditions of near autonomy at the local level and in the absence of commensurate power on the intermediate level of government, have been unable to adapt to new and changing demands that have significance beyond the local jurisdiction. The current structure has not been capable of sufficiently attuning (by means of intermunicipal or other arrangements) to aggregations of recipients of police services on other than geographical criteria.

(2) The coexistence of the National Police and the municipal forces has caused problems. A number of these problems can be explained by the fact that similarities between the two allow for comparisons in terms of payment, career structures, responsibilities, transferability, and the like. The similarities are deliberate; the local units of the National Police are organized so as to highly resemble municipal police, and they execute the same tasks. The dissimilarities are, of course, equally deliberate. They account for the fact that the comparisons often come out in favor of the municipal police. For instance, an officer has more possibilities to be promoted in municipal forces than in the National Police. Also National Police officers are often dissatisfied with sudden and relatively frequent transfers to other parts of the country. These transfers—temporary or permanent—occur much less often in the municipal police agencies, and the consent of the officer is generally required.

Another problem stems from the organizational structure of the National Police. The size of its local detachments is seen as too small, especially under conditions where a regular reserve potential is almost nonexistent. Most requests for temporary or prolonged assistance, either by other communities (irrespective of their having municipal police or not) or by central authorities, have to be met by releasing officers from their regular tasks. Understaffed units are rendered even more incapable of performing their normal tasks.

The fundamental recommendations aimed at overcoming these institutional difficulties include the following:

(1) There should be one type of police. The coexistence of two police organizations performing highly similar functions is problematic, and one type of police for each function—a conceivable alternative—is "undesirable" because maintaining public order and combating crime are "too closely interconnected"; separating those functions will result in "conflicts of competence and associated consequences" (Van Agt and Geertsema, 1972:7). Moreover, the establishment of two monofunctional police organizations would lead to the (unacceptable) elimination of "the only real unity that has existed for over 140 years within the Dutch police, namely the unity which, within the individual police officer, resolves in itself all contradictions" (Van Agt and Geertsema, 1972:7).

(2) The new police forces should have a sufficient strength (minimum 400 members) to permit effective and efficient performance, internal organization, training, task specialization, and use of modern equipment.

(3) The new forces should be embedded in a structure of regional government that can exercise commensurate democratic control over them.

(4) Both requirements 2 and 3 above can, and should[6] be met by linking the design for provincial police to current plans for national reform of the public administration.

(5) There should be intensive and permanent triangular consultation. This consultation should be institutionalized in order to:

- Partially compensate the reduced power that a mayor will have over the police
- Coordinate the demands which police will receive from the various officials having operational authority
- Coordinate rule-making authority among the various operational authorities
- Promote the societal relevance of police services.

(6) Legislators should refrain from deciding that one police task (i.e., the order task) has primacy over the other (the justicial). This has repeatedly been suggested by many institutions and organizations participating in the discussions on police reform. The justification which the ministers advance for this recommendation is that one (notably the Minister of Justice) should not have responsibility for something over which one has no control. The format of the core constitutional authority should be such that it lies with the ministers

separately for matters exclusively concerning their specific responsibilities and jointly for all other matters.

(7) There should be a National Council for the Police. Because police provide services for all citizens of the nation and to all administrative institutions at all levels, it can be expected that triangular consultation will not be sufficient to accommodate the differentiated demands and changing value orientations associated with public order and the means of achieving it. An integrative and coordinative function can be provided by a Council for the Police, constituted of representatives of the parties involved in the triangular consultation both on the provincial and the national level and representatives of various regional, professional, academic, industrial, and other segments of the population.

Additional recommendations pertain primarily to increased exceptional power over the police for the central authorities, the organization of centralized branches of the police providing highly specialized services.

Is there a consistent model of institutional design that underlies these recommendations? What are some of the effects that may be expected from the recommended structures?

It seems that the proposals reflect elements of two fundamental institutional models: the hierarchic and the democratic, or the centralized and the decentralized respectively. A majority of the central aspects of the new design are associated with the model of "democratic administration." They seem to be restricted, however, by considerations that relate to the hierarchical model of administration. Thus, for instance, the police will be regionally decentralized but their jurisdiction is mutually exclusive. This means that, apart from exceptional situations, the provincial police force will be the monopolized producer of police services within a region—from the point of view both of the public and of local and provincial authorities.

Conversely, there will be a system of partly overlapping operational authorities within the provincial structures (see Figure 2), but they are located at different levels of the governmental hierarchy. The possible consequence of the relatively weak influence of local authorities over the police is less responsiveness to local conditions. Commanding police officers can apply interpretations and priorities that are at variance with those of the local administration—especially when "it is up to the administrative authorities and the chiefs of police themselves to respond to the variety and multitude of demands" (Van Agt and DeGaay Fortman, 1975:5).

Finally, there is the assumption that an increase in the size of organizational units will provide a remedy to many of the ills of the prevailing system. The recommendation to consolidate is advanced with no other rationale than repeated reference to the general consensus on this issue. The attorneys general observed: "The thesis that the proposed strength [of the regional police] would be suitable for a sufficiently self-supporting and specialized police corps is not, to our knowledge, based on any solid research" (Procureurs-Generaal, 1973:7).

Unfortunately they do not contribute any additional argument to the thesis. Instead they fear: "that police forces of this size [reference here is to 500, RJV] will not be large enough to perform the multitude of general and specialized tasks in a satisfactory way" (Procureurs-Generaal, 1973:7). Indeed, the compelling "logic" of the situation—where a central objective is to abandon a dysfunctional dualism—seems to foreclose the logical possibility that perhaps not all dual or multiple need be dysfunctional.

An alternative line of reasoning might suggest that it is precisely the concentration of "multitudes of general and specialized tasks" within one organizational structure that contributes to the problems. On that assumption there would be a basis for evaluating functional segmentation. The presence of alternative solutions would also make it possible to formulate hypotheses about performance under different conditions. If these hypotheses were subsequently substantiated by empirical evidence and/or theoretical argumentation, a meaningful ground would exist for making choices from among an array of alternatives. In the current situation, alternatives are absent, and the proposed institutional design seems, in this respect, to be based on compulsion rather than choice.

Whatever the outcome of the research to which the attorneys general refer, it seems unlikely that it will indicate some optimal size or range for a single organization that has to perform many different functions. There is not yet a comprehensive theory that unambiguously describes the relation between organizational structure and function, but some aspects have been investigated.

Recent studies suggest, for instance, that it is questionable whether larger police agencies provide better police services to local communities than do smaller agencies.[7] Inherent in the rationale of arraying alternatives is the notion of costs that are incurred when opting for one or another institutional arrangement. It would not be too farfetched an expectation to find some reference to costs in the reform proposals. However, even the switch from the initially proposed size (400 to 500) to roughly double that (900) was made without any analysis of possible effects, such as diseconomies of scale for certain functions.

What might a design which takes into account the above considerations look like? Surely it is impossible to have as many police agencies as there are functions. Nevertheless, it seems feasible to divide the various functions into those requiring less centralized and smaller organizational units (such as the order task on a local level) and those requiring more centralized and larger units (such as the order task on a provincial or national level and the justicial task). Not only could such a dualism be expected to increase the quality of performance, it also would be compatible with the principle of local autonomy in matters that concern the local community alone. Why, one wonders, can a Minister of Justice not be held responsible for his command authority when he has no control over the police whereas a mayor is expected to be responsible under exactly the same conditions? A similar observation applies to the situation in which the attorneys general find themselves; they are, and will be even more so if the proposals are implemented, responsible for performance of the judicial task without having constitutional authority. This has led them to remark: "In the past few years we have experienced how the limited influence of the justicial branch of the administration on the determination of the strength of detective teams within the municipal police forces has had notable harmful consequences for the ... combating of criminality" (Procureurs-Generaal, 1973:4, 5).

It appears that these proposals contain incompatibilities concerning size, performance, and democratic control. It also appears as if these incompatibilities are sharpened by insistence on the principle of "unity of police" (in the police the order and justicial tasks are united), a principle that will be violated anyway given the fact that the centralized police agencies will provide specialized services.

In conclusion, it may be noted that many expected improvements pertain to the internal organization and operation of police agencies. Other expected improvements pertain to the quality of performance as reflected in crime rates, percentage of crimes cleared, and number of traffic accidents. None of the objectives, however, are stated in an operational way. Neither are there clear guidelines, in the proposals, for formulating measures of performance for police services, other than those relating to the justicial task. The scarce empirical evidence that is presented in the proposals bears primarily on the diagnostic assessment of the prevailing situation. The recommendations and design are not grounded on evidence, nor are the arguments always logically consistent, as has been indicated above. Logical consistency is, in itself, not a necessary condition for a "sound" design. Departures from logical consistency should, however, be justified in one way or another. When they are not, the design loses claim to the presumption that it is based on reasoned calculations.

Two issues are of crucial importance for the viability of the future police structure: First, will a commensurate democratic power actually be established on the provincial level—an issue that falls outside the Police Act proper? Second, will the organizational structure *within* a provincial police corps allow local authorities to assume democratic responsibility for the maintenance of order task in the community?

POSTSCRIPT

Assuming that "the security of the community"—the prime objective of police services—is affected by many factors (including population characteristics and such institutional arrangements as the court system and the welfare system), the Dutch situation offers an ideal opportunity for investigating how institutional arrangements contribute to that security and how changes in these arrangements relate to performance (see Ostrom, 1972:93-109). Almost all other factors remain relatively constant in view of the brief transitional period that has been suggested. Executing a study to test hypotheses on the relation between institutional design and performance may be justifiable in terms of advancing theoretical insight. There are, however, more practical reasons for undertaking the effort. Without such a study it may be extremely difficult, if not impossible, to evaluate the degree to which the effects of the changes correspond with the declared objectives and to determine what price is actually paid for and what benefits are realized from the new structure. Without such information additional adjustments will assume properties of ad hoc problem solving.

NOTES

1. See, for instance, the paragraph on "new developments" (Van Agt and Geertsema, 1972:5-7).

2. One example: "Threatening with, and implementation of, sanctions should be an ultimate means. Administrative intervention should come first. . . . Administrative intervention, prevention (of violations of the law), increasingly becomes more important. Meanwhile the character of justicial (repressive) intervention changes clearly" (Bizondere Commissie, 1973a). See also Punch, 1975; Weenk, 1974a, 1974b.

3. The initially suggested lower limit of 400 to 500 was apparently arrived at by dividing the total number of police officers by the 40 to 50 districts that were foreseen in reform proposals prior to the spring of 1975. At that time, plans for establishing 26 new style provinces were published and well-received.

4. This is the informal rule-making authority that an operational authority has through his organizational affiliation with a rule-making authority. An example would be the influence that a mayor in the proposed structure could have on rule-making through the Queen's Commissioner. See Figure 2.

5. The size of the regions currently served by the National Police is such that the number of municipalities within the jurisdiction of one group or section is relatively small. Hence the commanding police officers have to deal only with one or several mayors; they can, and do, identify themselves to a certain extent with local interests. In contrast, the future provincial chief of police will have to accommodate, in relation to local order tasks, an average of 33 mayors.

6. Even though it can be expected that the boundaries of the new provinces will not often, if at all, coincide with those defining the recipients of the services provided at that level, adequate democratic control is a decisive factor here. It is felt, for instance (P. van Dijke in a letter to the author, 1976), that arranging police forces on another basis than the new Structure of Provincial Movement, would be a severe hazard for democratic control over the police.

7. Examples are Ostrom and Parks, 1973; Ostrom, Parks, and Whitaker, 1973; Parks, 1971.

REFERENCES

Algemeen Politieblad (1975). "Regering kiest voor nieuwe indeling van Nederland in 26 provincies," pp. 413-420.

AGT, A. van, and DeGAAY FORTMAN, W.F. (1975a). "Memorie van Antwoord." In Tweede Kamer stukken, 10124 nr. 7.

——— (1975b). "Nauwere samenwerking rijks- en gemeentenpolitie in afwachting van gewestvorming." Pp. 207-208 in Algemeen Politieblad.

AGT, A. van, and GEERTSEMA, W.J. (1972). "Nota met vraagpunten." Pp. 3-11 in Tweede Kamer stukken, 10124 nr. 3.

Bizondere Commissie Herziening Politiewet (1973a). "Verslag van een gehoor." In Tweede Kamer stukken, 10124 nr. 5.

——— (1973b). "Voorlopig verslag." In Tweede Kamer stukken, 10124 nr. 6.

De Nederlandse Gemeente (1973a). "Beheer van politie hoort thuis bij organen van binnenlands bestuur," pp. 199-200.

——— (1973b). "Reorganisatie van politie moet op bestuurlijke gewestvorming vooruit-lopen," p. 201.

DIJKE, P. van (1973). "Gewestelijke politie hoeft niet op gewestwet te wachten (interview by F.J. van der Heijden). Pp. 37-40 in De Nederlandse Gemeente.

KUITENBROUWER, F. (1974). "De strijd om het politiebestel." Intermediar, 38(September):15-21.

Ministerie van Justitie (1974a). Samenwerking tussen rijkspolitie en gemeentepolitie. Mededelingen van het stafbureau voorlichtin nr. 848.

––– (1975b). Herziening politiewet. Mededelingen van het stafbureau voorlichting nr. 884.

MOLENAAR, A. (1974). "Op basis van tussenoplossing binnen 5 jaar gewestpolitie." De Nederlanse Gemeente, 38(September):453-454.

NEAVES, L. (1975). "Police reform proposition." Unpublished paper, Department of Poltical Science, Indiana University.

OSTROM, E. (1973). "On the meaning and measurement of output and efficiency in the provision of urban police services." Journal of Criminal Justice, 1:93-112.

––– (1975). "Size and performance in a federal system." Unpublished research paper, Workshop in Political Theory and Policy Analysis, Indiana University.

OSTROM, E., BAUGH, W.H., GUARASCI, R., PARKS, R.B., and WHITAKER, G.P. (1973). Community organization and the provision of police services (Sage Professional Papers in Administrative and Policy Studies, 03-001). Beverly Hills, Calif.: Sage.

OSTROM, E., and PARKS, R.B. (1973). "Suburban police departments: Too many and too small?" Pp. 367-402 in L.H. Masotti and J.K. Hadden (eds.), The urbanization of the suburbs. Beverly Hills, Calif.: Sage.

OSTROM, E., PARKS, R.B., and WHITAKER, G.P. (1973). "Do we really want to consolidate urban police forces? A reappraisal of some old assertions." Public Administration Review, 33(September-October):423-433.

PARKS, R.B. (1971). "Measurement of performance in the police sector: A case study of the Indianapolis police department." Unpublished manuscript, Workshop in Political Theory and Policy Analysis, Indiana University.

POP, J.J.H. (1973). "Kamercommissie in Voorlopig Verslag: 'Gewestpolitie moet er komen en gauw ook!' " De Nederlandse Gemeente, 41(October):477-479.

––– (1974a). "Van rijkspolitiesamenwerking naar gewestpolitie." De Nederlandse Gemeente, 38(September):457-458.

––– (1974b). "Conclusie op Symposium Rijkspolitie: Geen nationale politie." De Nederlandse Gemeente, 42(October):508.

––– (1974c). "Vooruitlopend op gewestpolitie: Samenwerking van rijks- en gemeente-politie op komst." De Nederlandse Gemeente, 47(November):564-565.

Procureurs-Generaal bij de Gerechtshoven (1973). Bijlage IV. Jaarverslag Openbaar Ministerie, The Hague.

––– (1974). Bijlage IV. Jaarverslag Openbaar Ministerie, The Hague.

PUNCH, M. (1975). "Rayonagent: Politieagent also maatschappelijk werker." Algemeen Politieblad, pp. 51-54.

REENEN, P. van, and VERTON, P.C. (1974). "Over do legitimiteit van het hedendaagse politie-optreden." Mens en Maatschappij, 1(January):72-78.

SCHOLS, A. (1974). "Gemeentepolitie heeft voordelen en grenzen." De Nederlandse Gemeente, 38(September):459.

Van BANNING, J. (1974). "Het slepende probleem van de politie-organisatie." De Nederlandse Gemeente, 46(November):556.

Vereniging Nederlandse Gemeenten (1972). In Tweede Kamer stukken, 10124 nr. 3.

––– (1973). In Tweede Kamer stukken, 10124 nr. 5.

VOS, H.M. (1975). "De politie als middel van sociale beheersing." Algemeen Politieblad, pp. 99-103.

WEENK, E. (1974a). "De wijkagent/rayonagent (I)." Algemeen Politieblad, pp. 3-8.

––– (1974b). "De wijkagent/rayonagent (slot)." Algemeen Politieblad, pp. 27-34.

Part IV

ASSESSING PERFORMANCE

8

Public Versus Private
Production Efficiency in Switzerland:
A Theoretical and Empirical Comparison

WERNER W. POMMEREHNE
BRUNO S. FREY

□ IN THIS ESSAY *public* (municipal) production of an urban service is compared to production of the same service by *private firms*. Such a comparison must be based upon a theoretical analysis in order to derive meaningful conclusions from empirical evidence.

The purpose of this essay is to:

- make a contribution to a research area in which economics and public policy analysis have had *little application.*[1]

- test propositions about the *relative efficiency* of governmental versus private production, thus stressing the comparative analysis of institutions.

- make some suggestions about *desirable organizational arrangements* for a specific municipal service—namely, residential refuse collection.

AUTHORS' NOTE: *The authors are indebted to many people in industry and government of various Swiss cities for granting interviews and responding to mailed questionnaires, as well as for providing the basic data for this study. They are particularly grateful for the useful advice provided by U. Geissmann of the Union of Swiss Cities, and R. Roth of the Swiss Federal Office of Statistics and the valuable comments of F.P. Bish of Indiana University.*

The emphasis lies in a comparative approach to both theoretical analysis and empirical research; no institutional arrangement is perfect in the sense of achieving completely efficient (cost-minimizing) production. If this basic view is accepted, it follows that it will be necessary to analyze and contrast institutions which are *imperfect*. "Market failure" and "governmental failure" coexist.

In this paper, we first consider the arguments for and against private and public production. There are valid theoretical arguments for and against each kind of production arrangement. Empirical evidence is necessary to test the relative efficiency of the two arrangements. In the second part, a model of refuse collection for Swiss cities is developed and empirically tested.

On the basis of this analysis, it is concluded that, in the cities studied, private is more efficient than governmental production. On the other hand, as argued in the final part of this paper, the governmental task does not end if the service is privately produced. Governmental activity instead takes a new form: that is to develop safeguards to assure that private production will be more efficient. In particular, it needs to find ways to maintain competitive pressures. This involves problems of information, uncertainty, and control.

THE RELATIVE EFFICIENCY OF PUBLIC AND PRIVATE REFUSE COLLECTION: THEORETICAL CONSIDERATIONS

THE NATURE OF REFUSE COLLECTION

Refuse collection has (practically) no properties of a collective good, compared to, for example, public transport or schools. There are, however, negative externalities connected with *consumption,* namely the bad smell and public health hazards resulting from untreated wastes (Gueron, 1972). There has, therefore, been public intervention in the production of this service for a very long time, It is *collectively* decided that wastes must be disposed of in regular sequence and treated at specified locations.

There are few significant externalities associated with the *production* of refuse collection (if the noise involved in picking up garbage is disregarded). Although final disposal of wastes may be subject to strong negative external effects, this paper is concerned only with refuse *collection*. It is sufficient to indicate that final waste treatment in Switzerland usually rests in the hands of the cantons (states) or the central government. The lack of significant *external-ities of production* connected with refuse collection means that a central reason for governmental production does not apply. There is no theoretical reason why there should not be private production under competitive conditions.[2]

PITFALLS IN PUBLIC PRODUCTION

There are two main arguments against public production:

(a) The fact that there are public property rights in the production unit means that there is more discretion by the managers, enabling them to run the firm less efficiently.

(b) A publicly run production unit is subject to more restrictions than a private firm because it is part of a general system of public administration. These additional restrictions are very likely to impair production efficiency.

These two arguments are discussed in more detail below: the first section relates to the property rights aspect (see, e.g., Alchian, 1967, and Furubotn and Pejovich, 1972) and the second to bureaucratic organization.

Inefficiency Due to Public Ownership

Managers of a *private* enterprise are, under perfectly competitive conditions, forced to follow the interests of the shareholders, i.e., to maximize profits and the value of the firm in terms of share prices. If an inefficient policy is undertaken, the present owners will interfere, or the resulting reduction in the value of shares will enable outsiders to gain the majority. In both cases, the inefficient management will in theory be dismissed. Therefore it is in the managers' own interest to choose an optimal output (i.e., to take into account consumers' preferences) and to produce services at the lowest cost possible. Under conditions of perfect competition, the quality and price of the product are continually controlled by the market.

Control of managerial behavior is much weaker in *public* enterprises. The taxpayer, as the "owner" of the firm, and the consumer, as the user of the product, have little incentive to exert effective control. To evaluate how efficiently a public "firm" is run, individuals must be willing to bear the costs associated with obtaining information about the output and costs of that firm. The costs of obtaining this information may be very high—particularly where the goods themselves are not amenable to measurement and evaluation. Given the uncertainty of achieving increased efficiency, and the small benefit likely to accrue to an individual as a result of such improvements, most individuals would find it irrational to devote much time and/or effort to such activity.

With public production, the consumer rarely has the choice to react as he might in the market, namely to "exit" (see Hirschman, 1970:21 ff.), i.e., to switch to a competitive supplier if the product is unsatisfactory. Often, the only possibilities may be to boycott the service altogether or to move to another community. Both of these alternatives may involve sizable costs.

"Voice" or protest is another means of influencing the quality and efficiency of public service delivery. It is, however, likely to be effective discontinuously, only. Furthermore, in the special case of refuse collection, it is unlikely to be a focal point around which professional voice-makers (the opposition party) can organize a successful election campaign. This may happen only if obvious and large mistakes are being made. In theory, management will avoid undertaking actions which might lead to such mistakes. However, there is still much room for production inefficiencies.

Inefficiency Due to Bureaucratic Organization

Inefficiencies in municipal production of public services may also result from the manner in which the public bureaucracy is organized. Bureaucratic inefficiencies result from a variety of factors.

Often the goods and services offered by a public firm are financed through the general budget or by various kinds of taxes not directly related to the quantity consumed. This arrangement may provide a *positive* incentive to managerial inefficiency, particularly if salary and prestige are linked to budget size, i.e., to the costs of operation. Bureaucratic *rules* and *regulations* may have an even stronger impact—affecting both the input and output side of a public enterprise. With respect to *inputs*, it is reasonable to assume that trade union influence is stronger and likely to affect productivity more negatively in public than in private enterprises. Politicians often depend on the trade union vote for survival and may be inclined to grant wage demands and other favors, because the costs of such actions are distributed broadly over a large section of the population.

Another reason why politicians may be reluctant to oppose trade union demands is that strikes may strongly reduce their reelection chances.

Public managers usually have little say in their sector's wage determination. They may not be able to adjust relative wages to reflect the marginal productivity of various occupations and workers. They have little opportunity to set wages to give positive incentives for better work. Generally there are rigid constraints on wage differentiation. Wage increases must be set following bureaucratic rules—in particular, regarding age and seniority. The restrictions imposed with respect to wages are marked in the case of refuse collection where 60% to 80% of costs consist of wages.[3] Managers of municipal enterprises are also subject to strong restrictions with respect to hiring and firing. Often workers must be kept on the payroll who, in the private sector, would have been dismissed.

Input restrictions of the kind mentioned are not absent in private firms, especially in large enterprises. However, what matters is that such constraints are much *greater* in the public sector and that managers of private firms have more scope for differentiation and discrimination (see Becker, 1971:31 ff.; Arrow, 1973:10).

There are also bureaucratic restrictions on the *output* side. The production level and—more important from the point of view of costs—the distribution among the various kinds of output are often governed by rules and regulations. The management has few opportunities to vary the composition of supply in response to changes in costs and (perceived) demand conditions.

There are thus a number of a priori theoretical reasons why public production may be "inefficient." It may appear to be advisable to switch from public to private production. This conclusion is, however, warranted only if private production is organized so as to guarantee the efficiency properties pertaining to the model of competitive supply. The point of reference must be the functioning of an institution in the real world, and not a model in which "perfect efficiency" is guaranteed by the appropriate choice of *assumptions.*[4] The following section is concerned with the possible inefficiencies of private production.

INEFFICIENCY IN THE PRIVATE SECTOR

In the private market economy, inefficiencies may arise from several kinds of *lack of competition.*

In the case of residential refuse collection, competition does not occur among a large number of firms.[5] Private collectors receive a license allowing them to operate in the market, usually under a contract for a fixed period of time. Even if there are several firms in the market, there are certainly not the large number of price takers required to meet the assumptions of the model of perfect competition. Competition is, at best, oligopolistic, making possible a wide variety of results with respect to efficiency.

Market entrance is not free. Although start up costs in residential refuse collection are not high, a production license must be obtained and there are many rules and regulations which must be observed.[6] Licenses may be acquired by buying out an existing firm (an option which rarely occurs) or as a result of a decision by authorities that additional suppliers are needed to increase the level and/or quality of performance.

Firms already in the market may, however, oppose such an extension. It may be worthwhile for them to form an *interest group* which seeks to restrict entry, to divide the collection area among existing firms, and to set the rates to be requested from the various local authorities. These prices may be set at a level where the most inefficient firms can survive, providing all other firms with a *differential rent.*

These rates often have the character of a cost-plus arrangement. Such cost-plus arrangements may encourage firms to tolerate cost increases because they will be compensated anyway. Under these circumstances—as in the public sector—private firms cannot be expected to resist excessive wage demands.

Such regional monopolies or oligopolies are likely to develop as a result of the structure of incentives facing firms already in the field. The problem cannot be solved merely by suggesting that such a development "should not be allowed to happen." The formation of imperfect markets is the result of a socio-political process within which there are actors operating in their own self-interest. This process cannot be changed simply by invoking the overall superiority of perfect competition. It may well be that those involved in refuse collection, including both private firms and public bureaucracies, find such solutions more advantageous than increased competition. Private contractors and the public bureaus which are supposed to fix the licenses and contracts so as to achieve maximum efficiency often form a tacit coalition based on their mutual dependence and long acquaintance (Stigler, 1971).

INTERIM CONCLUSIONS

The case for the superior efficiency of private production is no longer so obvious as when public sector inefficiencies are discussed in isolation. There are certainly many convincing reasons why public production is likely to be inefficient. It has, however, been shown that when the assumption of perfect competition is abandoned, inefficiencies must also be expected in the private sector.

Theoretical reasoning alone cannot settle the dispute of whether private or public production is *more efficient*. It is necessary to resort to *empirical tests.*

THE RELATIVE EFFICIENCY OF PUBLIC AND PRIVATE REFUSE COLLECTION: EMPIRICAL ANALYSIS

PREVIOUS STUDIES

The simplest method of determining efficiency differences between private and public production is to calculate the *per-unit* costs of refuse collection, the unit being the household, or the quantity or weight of waste. A comparison using this measure alone is, however, most unsatisfactory because differences in the *quality* of service (e.g., where the waste is picked up) and in *technical conditions of production* (e.g., whether the topography is hilly) are not allowed for.

A model which takes these differences into account has been devised (Hirsch, 1965). It is assumed that the cost of refuse collection depends on the amount of service (U), the quality of service (Q), the technical conditions of production (T), the factor prices (FP), and the level of technological knowledge (TK). The average unit cost per ton (AC) thus is:

$$AC = f(U, Q, T, FP, TK, D_1, D_2).\qquad(1)$$

The function also includes two dummy variables (D_1 and D_2) which represent the institutional condition of production (private versus public) and the type of financing (charges to private households versus financing through the general budget).

In the estimation equation of household refuse collection in 24 suburbs of the St. Louis City-County area the following specific variables were included (Hirsch, 1965:90 ff.):

X_1 = 1960 average annual residential refuse collection and disposal cost per pickup in dollars.

X_2 = number of pickup units.

X_3 = weekly collection frequency.

X_4 = pickup location, where curb pickup is 0 and rear of house pickup is 1.

X_5 = pickup density, i.e., number of residential pickups per square mile.

D_1 = nature of contractual arrangements, where municipal collection is 0 and private collection is 1.

D_2 = type of financing, where general revenue financing is 0 and user charge financing is 1.

The estimation yields:

$$X_1 = 6.16 + 0.000\ 089\ X_2 - 0.000\ 000\ 000\ 436\ X_2^2 + 3.61^* X_3 + 3.97^* X_4$$
$$\quad\quad (0.000\ 195)\quad (0.000\ 000\ 000\ 832)\quad (1.14)\quad (1.50)$$

$$\quad - 0.000\ 611\ X_5 - 1.87\ D_1 + 3.43^* D_2$$
$$\quad\ (0.000\ 442)\quad (2.40)\quad (1.10) \hspace{4cm} (2)$$

$$\bar{R}^2 = 0.76; N = 24.$$

The standard deviations are given in parentheses below the respective parameter estimates, which have an asterisk if they are statistically significant at the 95% confidence level.

It should be noted that the model tests the costs of waste collection *and* disposal and that the costs refer to the *costs per pickup* (and not per household or per ton). The dummy variable D_1 for the type of production institution had no statistically significant influence upon costs; it seems that there is no clear difference of efficiency between private and public production. The type of financing exerts a significant influence on costs. It is, however, rather surprising that the coefficient is positive. It would be expected a priori that charges (which are assumed to be similar to prices) would lead to a higher cost-consciousness on the part of consumers and producers, leading to a pressure to hold costs down.

The relative efficiency of different institutional forms of production of refuse collection was also tested recently for 27 public and private enterprises in Montana (Pier, Vernon, and Wicks, 1974). A production function with fixed proportions was found to fit the data best. The cost functions derived led to the conclusion that public production was less efficient than private production at a low output level and more efficient at a high output level when *capital costs* are considered. With respect to *labor costs* public production was found to be more efficient than private across all size ranges. For localities larger than about 1,750 inhabitants, average costs for municipal refuse collection were lower than for private firms.

The Montana study is interesting, particularly because it attempts to estimate a production function. It must, however, be criticized on various grounds.[7] Output is not measured by quantities or the number of households served, but rather by pickup places. Differences in the quality and technical conditions of refuse collection are not sufficiently allowed for. Moreover, the methodological approach used is inappropriate for examining the relative efficiency of institutions. *Separate* production and cost functions were estimated for private and public production. A comparison of the result with respect to efficiency is admissible only if the two sectors produce under the same general conditions. In general this cannot be expected. To test the relative efficiency of the two institutional arrangements *one single* estimation equation is required for private and public production. The relative effect on average costs then can be isolated by introducing dummy variables for the two types of service institutions.

A MODEL OF REFUSE COLLECTION FOR SWISS CITIES

A cost function of residential refuse collection in 103 Swiss cities was estimated with the intention of avoiding the shortcomings of the previous studies. About half of these cities are served by private collectors (mostly on a contract basis), and the other half by public enterprises.

Average costs of refuse collection per residential household (AC) are assumed to be influenced by quantitative and qualitative factors, by the technical conditions of production, by factor prices and the state of technology. The *quantity* of refuse collected is measured by the average annual number of tons of refuse carried away from each household (U_1). In Switzerland it is also important to consider the additional refuse created by the inflow of foreign workers in the summer and/or winter (U_2) and, in a smaller degree, the number of tourists lodging in private apartments and homes (U_3). Another important factor is the refuse that may be generated by individuals commuting between place of residence and place of work. Thus U_1 is amended by three additional variables: U_2 is the annual average number of foreign workers, relative to the residential population; U_3 is approximated by the average annual number of lodgings for night, relative to the residential population; and U_4 measures the decrease in household refuse collection demand due to outgoing commuters, relative to the residential population.[8] It is expected a priori that U_1 to U_3 (U_4) tend to increase (decrease) average cost.

The *quality* indicators consist of pickup location (Q_1), number of collections per week (Q_2), and the distinction between joint ($Q_3 = 1$) or separated ($Q_3 = 0$) collection of ordinary household garbage and bulky wastes (Q_3). Average costs are expected to rise the farther from the street the refuse is picked-up and the more often it is collected (because the quantity per collection is reduced). The effects of collecting bulky wastes along with other household refuse cannot be determined a priori. Although the collectors are already at the pickup point, the handling is more difficult and there is less possibility for automation. Other factors affecting quality are the noise produced (Q_4) and the mess left on the street (Q_5), which are, however, difficult to determine empirically.[9]

Among the *technical conditions* of refuse collection are the economies connected with increasing density of collection area, which is split up into the number of pickup points per kilometer of city streets (T_1) and the number of households per pickup point (T_2). Another important feature in Switzerland is the topography, measured by the differences in height within a city (T_3) and the intensity of snowfall, approximated by the number of days with snowfall (T_4). The larger these differences and the more intensive the snowfall, the higher are time and fuel costs. Finally, the distance to the waste deposit site (T_5) may be expected to positively affect average costs.

The *factor prices* (FP) have an influence on the input mix; the higher the relative cost of labor, the more capital will be substituted for labor.

The *state of technological knowledge* and its precise influence on average costs is difficult to evaluate independent of factor prices. It may well be that the crew size (TK_1) and type of equipment (TK_2) are determined by relative factor prices and do not have a direct influence on average costs. If, however, there are institutional constraints making it impossible for a refuse collection firm to take advantage of new technology, costs may be directly affected (see Downing, 1975:9 ff.). The same argument applies, of course, to the firm's capacity to adapt to changes in relative factor prices.

Inclusion of these variables allows one to control all relevant factors and thus to compare average costs for *public* ($D_1 = 0$) and *private* ($D_1 = 1$) production. The *type of financing* is controlled through another dummy variable D_2 ($D_2 = 0$ if general budget financing, $D_2 = 1$ if the households are charged).

The estimation equation thus is:

$$AC = h\ (\underbrace{U_1, U_2, U_3, U_4}_{\text{quantity}};\ \underbrace{Q_1, Q_2, Q_3, Q_4, Q_5}_{\text{quality}};\ \underbrace{T_1, T_2, T_3, T_4, T_5}_{\text{technical conditions}};$$

$$\underbrace{FP}_{\substack{\text{factor}\\\text{prices}}};\quad \underbrace{TK_1, TK_2}_{\substack{\text{technological}\\\text{knowledge}}};\quad \underbrace{D_1, D_2}_{\substack{\text{institutional arrangement}\\\text{and type of financing}}}).\qquad (3)$$

The theoretically expected signs of the partial derivatives are:

$$\frac{\partial h}{\partial U_1}, \frac{\partial h}{\partial U_2}, \frac{\partial h}{\partial U_3},\qquad \text{all} > 0;$$

$$\frac{\partial h}{\partial U_4},\qquad\qquad < 0;$$

$$\frac{\partial h}{\partial Q_1}, \frac{\partial h}{\partial Q_2},\qquad \text{both} > 0;$$

$$\frac{\partial h}{\partial Q_4}, \frac{\partial h}{\partial Q_5},\qquad \text{both} < 0;$$

$$\frac{\partial h}{\partial T_1}, \frac{\partial h}{\partial T_2},\qquad \text{both} < 0;$$

$$\frac{\partial h}{\partial T_3}, \frac{\partial h}{\partial T_4}, \frac{\partial h}{\partial T_5},\qquad \text{all} > 0;$$

$$\frac{\partial h}{\partial D_2} < 0;$$

The other signs cannot be determined through a priori reasoning.

APPLICATION OF THE MODEL

An empirical application of this model was undertaken using data collected by the authors on refuse collection in the 112 largest Swiss cities. The smallest city included in the survey was Stans, with 5,100 inhabitants, and the largest was Zürich, with 422,600 inhabitants. More than half of the Swiss population lives in these 112 cities.

The data on costs, which include operation costs, depreciation and interest on the capital invested, and other data relevant to equation (3), were collected by questionnaire. Additional data were taken from the Statistics of Swiss Cities and from the results of the population census of 1970.[10] All the data refer to 1970.

Cities in which industrial and household refuse are collected jointly were excluded. Not all of the necessary data were available for a number of other cities. For the estimation of the average cost curve, there remained 103 cities, of which 55 have a municipal and 48 a private refuse collection service. For some of the variables no satisfactory data could be obtained, in particular for factor prices (FP) and for the state of technological knowledge (TK_2).

The estimation results using the remaining variables are given in Table 1. Standard step-wise least squares multiple regression was used, leaving only those variables in the equation which were statistically significant at the 95% confidence level, or nearly so. A look at the correlation matrix indicates that there are no serious problems of multicolinearity.[11] The variables for outgoing commuters (U_4) and pickup location (Q_1) and for the joint or separate collection of household garbage and bulky wastes (Q_3) have been excluded due to their insignificant effect on average costs. This result is quite plausible: in Switzerland, refuse is almost always placed near the street so that there are too few observations for the alternative case. The opposing effects on average costs connected with Q_3 just seem to balance.

Equations (4) and (5) in Table 1 differ in that they are designed to answer different questions. In equation (4) the problem of the *budgetary implications* for the households and taxpayers is studied. Accordingly, a ceteris paribus comparison between average *costs* of municipal production (assuming near-zero profits) and *prices* of private suppliers is undertaken.[12] This does not constitute a comparison of production efficiency.[13]

In equation (5) the question of *relative production* efficiency is studied. A comparison is made between average real production costs of public and private enterprises. Thus, in the case of private firms, the average gross profits before tax are deducted. It is assumed that the average gross profit rate of private suppliers amounts to a markup of 7% upon production costs.[14]

All variables (with the exception of D_2) have the theoretically expected sign. The most important result is that the statistically significant dummy variable D_1, indicating the institutional form of production, has a negative sign: *public production* of refuse collection seems to be subject to *higher average costs* than private production.[15] This result does not change when the five largest Swiss cities (Zürich, Basle, Geneva, Berne, and Lausanne), which all have a municipal refuse collection, are excluded.

The sign of the dummy variable standing for the type of financing, D_2, is positive (although at a somewhat lower degree of significance), which is contrary to a priori theoretical expectations, but it corresponds to Hirsch's and Kitchen's estimates. This sign is plausible if household user charges are in no direct relation to production costs and to the extent of use by each household and/or if the additional costs due to fixing and implementing of the charges are large.

In order to test whether there are economies of scale in production, the *cost per ton of refuse collected* (instead of per household of the residential population as in Table 1) is taken as the dependent variable. Estimates with a log-linear specification proved superior, and they are presented in Table 2. Equation (6) gives the result when costs are compared to prices; equation (7) when real costs are considered only.

With the exception of U_2 and Q_2, all coefficients are statistically significant, and all—with the exception of U_1—have the same sign as in Table 1. The negative sign and the size of the coefficient of U_1 indicate that there are substantial *economies of scale in production:* average costs per ton of refuse collected decrease the more refuse that is collected.

This result contradicts those results reached in comparable studies for some districts in the United States.[16] It is, however, compatible with the results of the more comprehensive study by Savas (1976) covering the whole United States. He concludes that with increasing size of the American cities there are significant economies of scale in production, that the economies are strongly decreasing with an approximate city size of 30,000 inhabitants, and that from 50,000 inhabitants on there are neither positive nor negative economies of scale. This seems to be similar for Switzerland: if cities over 100,000

TABLE 1

ESTIMATES FOR REFUSE COLLECTION COSTS PER HOUSEHOLD (in sFr), 103 SWISS CITIES, 1970 (Linear Specification)

Eq.	Constant	Quantity			Quality	Technical Conditions					Institution	Financing	Test Statistics	
		U_1	U_2	U_3	Q_2	T_1	T_2	T_3	T_4	T_5	D_1	D_2	$\bar{R}^2$	F
(4)	21.19 (1.92)	7.47** (2.50)	4.73* (1.73)	2.30** (6.61)	9.34* (2.01)	-0.34* (-2.15)	-5.04* (-1.97)	0.17* (2.31)	0.23* (2.25)	1.23* (3.12)	-6.57* (-1.92)	5.96 (1.60)	0.65	18.25
(5)	19.06 (1.79)	7.16** (2.49)	4.67* (1.78)	2.20** (6.55)	9.11* (2.03)	-0.32* (-2.09)	-4.82* (-1.95)	0.17* (2.29)	0.21* (2.17)	1.16** (3.07)	-9.03** (-2.75)	5.57 (1.53)	0.65	18.16

U_1 = household refuse (including bulky wastes) per residential household (in tons)
U_2 = inflow of seasonal workers (in percent of residential population)
U_3 = inflow of privately lodged tourists (average daily number of lodgings for the night in relation to the residential population)
Q_2 = frequency of refuse collection per week: twice Q_2 = 0, more than twice Q_2 = 1
T_1 = number of pickup points per kilometer city street
T_2 = number of households per pickup point
T_3 = differences of height within the city region (in meters)
T_4 = average number of days with snowfall
T_5 = distance between the center of the locality and final refuse disposal site (in kilometers)
D_1 = institutional production conditions: public production D_1 = 0, private production D_1 = 1
D_2 = type of financing: if largely financed by the general budget D_2 = 0, if largely financed by charges to the households D_2 = 1

The figures in parentheses below the parameter estimates indicate the t-values. An asterisk indicates statistical significance at the 95% level, two asterisks at the 99% level of security.

TABLE 2

ESTIMATES FOR REFUSE COLLECTION COSTS PER TON (in sFr), 103 SWISS CITIES, 1970 (Log-linear Specification)

Eq.	Constant	Quantity			Quality	Technical Conditions					Institution	Financing	Test Statistics	
		U_1	U_2	U_3	Q_2	T_1	T_2	T_3	T_4	T_5	D_1	D_2	$\bar{R}^2$	F
(6)	1.17** (6.60)	-0.75** (-9.57)	0.09 (0.80)	0.10** (3.85)	0.06 (1.51)	-0.73** (-3.41)	-0.69** (-3.08)	0.79** (3.77)	0.22** (2.87)	0.18** (3.80)	-0.09** (-2.79)	0.08** (2.47)	0.62	4.82
(7)	1.14** (6.38)	-0.75** (-9.47)	0.09 (0.83)	0.10** (3.76)	0.06 (1.53)	-0.74** (-3.43)	-0.70** (-3.08)	0.79** (3.79)	0.21** (2.82)	0.18** (3.77)	-0.13** (-4.16)	0.08** (2.45)	0.64	5.32

For notes see table 1

TABLE 3

ESTIMATES FOR REFUSE COLLECTION COSTS PER TON (in sFr), 103 SWISS CITIES, 1970 (Log-linear Specification)

Eq.	Constant	Quantity				Quality	Technical Conditions					Institution	Financing	Test Statistics	
		U_5	U_5^2	U_2	U_3	Q_2	T_1	T_2	T_3	T_4	T_5	D_1	D_2	$\bar{R}^2$	F
(8)	1.05** (4.38)	-0.87** (-4.37)	0.33** (3.30)	-0.09 (-0.60)	0.06* (1.82)	0.10* (1.72)	-0.45 (-1.58)	-0.17 (-0.50)	0.62* (2.18)	0.14 (1.32)	0.30** (4.68)	-0.14** (-3.22)	0.06 (1.23)	0.43	7.56
(9)	1.03** (4.42)	-0.82** (-4.26)	0.31** (3.21)	-0.07 (-0.49)	0.05 (1.58)	0.10* (1.84)	-0.36 (-1.29)	-0.03 (-0.10)	0.49* (1.78)	0.12 (1.24)	0.30** (4.83)	-0.19** (-4.47)	0.05 (1.01)	0.45	8.05

U_5 = absolute weight of household refuse (in 1000 tons)

U_5^2 = square of M_5 (in 10^6 tons)

For other notes see table 1

inhabitants are excluded, the coefficient for U_1 is even more highly statistically significant.

It may be argued that inhabitants (Kitchen, 1976; Savas, 1976; Quigley and Kemper, 1976), the number of pickup units (Hirsch, 1965), or the average volume of refuse per household (Downing, 1975; see also our estimates in Table 2) are inadequate measures of output. To study the problem one should use the *absolute* weight or volume of refuse. This is done in Table 3, where the absolute weight of waste for each city (U_5) and its square (U_5^2) are included, the latter in order to take into account possible nonlinear relationships. Equation (8) compares costs and prices, equation (9) only costs.

The highly significant sign for the coefficient of U_5 and the positive sign for U_5^2 indicate that the conclusions so far reached are correct: the increase in total household waste is, over a wide range, accompanied by decreasing average costs which rise only at a very high tonnage. The positive *economies of density* reflected by the negative coefficients of T_1 and T_2 in Tables 1 and 2 are no longer visible in the new estimates.

It should be noted that, in all equations (4) to (9), the dummy variable for institutional arrangement of production is statistically significant and of the same sign.

CONCLUSIONS

From the results of this empirical analysis it appears that private production of refuse collection may be preferable to public production. Separation of demand articulation and production has been suggested as one means of overcoming some of the bureaucratic inefficiencies discussed earlier.[17]

As argued above, however, private production will not necessarily be more efficient than public production if imperfect market conditions exist. Some goods and services are publicly *produced* because of a concern that private firms will form coalitions and will be successful in restricting competition. The existence of sizable economies of scale in production, suggested by the empirical analysis, is another factor encouraging the establishment of monopolistic markets (Bish and Warren, 1972).

The possible existence of economies of density points in the same direction. In an area with low density (as measured by the number of pickup points per kilometer city street and the number of households per pickup point) there are unfavorable technical conditions for each producer, if several of them are producing. From the point of view of the producer it may be desirable to split the area, so that in each area there is only one supplier. This means, however, that there is no competition—at least for the time period in which a license is granted by the authorities.

Obviously, *the governmental task does not end* if it is decided that refuse collection should be undertaken by private enterprises. On the contrary, it is important that governments set *conditions* which make private producers function efficiently over the long run. They have to find ways and means to issue licenses and contracts stimulating competition and cost reductions. If governments provide the wrong incentives for private firms, nothing is gained by leading production to the "market."

The contracts offered may, for example, take the following form (see also Young, 1974):

(1) They should not cover too long a period (4-6 years) in order to stimulate competition for the production potential.

(2) The beginning and end of contract periods should be different from one city to another in order to give firms continuous possibilities to enter the competition for contracts.

(3) To minimize the risk of bad contracting, "performance bonds" should be required, so that the financing institution is responsible (up to the amount of the bond) for undertaking collection where bonded contractors fail to abide by the terms of the contract.[18]

In *low density areas* it is possible to produce pressure to keep costs down by establishing *competitions for the contracts*. In *high density areas* there is more scope for open competition in overlapping and even in identical markets. One may also think of establishing nonprofit enterprises designed to compete with private suppliers. For the reasons discussed above, one should not expect too much from such an arrangement: it may be that managers of such nonprofit institutions are also inclined to highly value a quiet life and to enter into tacit agreements with private suppliers and government bureaucracy. There is no easy way to guarantee competition in markets in which there are factors making collusion worthwhile.

SUMMARY

The relative efficiency of private versus public production of residential refuse collection is studied using the theory of property rights and the economic theory of bureaucracy. The empirical analysis of refuse collection in 103 Swiss cities suggests that private (contract) production may be more efficient. If refuse collection is turned over to private enterprise on the basis of some such evidence, the governmental task, however, does not end. Rather, its role becomes one of establishing the conditions necessary to ensure that private producers function efficiently in the long run. In particular, government must ensure that competitive pressures persist, if the potential savings from private production are to be realized.

NOTES

1. Exceptions which are concerned with production and cost functions of public services and which take account of the specific problems of output measurement in this connection are, e.g., E. Ostrom (1973) and Emerson (1975) on police, Ahlbrandt (1973) on fire protection, and several studies on schooling, surveyed by Hirsch (1973, chap. 11).

2. If there are (sizable) production externalities the aggregate production possibility set may become nonconvex, with the result that the competitive price system can no longer be trusted to bring about the (Pareto) optimal output level. If, however, the optimal production level is known through some other decision-making mechanism, it may still be advantageous to *set* the relative prices so that competitive firms attain it efficiently. For the whole problem see Baumol and Oates (1975, chap. 8).

3. This figure applies to Switzerland. In the United States and Canada it is of similar magnitude; see U.S. Department of Health, Education and Welfare (1970) and Kitchen (1976).

4. This criticism must be raised against many analyses in the tradition of the theory of property rights where all too often a perfectly functioning competitive private firm is contrasted with the inefficiencies of public organization. The opposite mistake is committed even more often. It is often said that if there is market failure, the government must take over the corresponding activity (but see McKean and Browning, 1975).

5. The following exposition is more general, but also applicable to Switzerland. Here, the main difference from other countries—such as, for example, Germany—lies in the conditions of licensing and contracting. The situation is somewhat different for industrial refuse collection.

6. Especially in Germany, the legislation is so difficult to understand that small firms which cannot afford a lawyer may be unable to enter a market. This tendency has increased since the introduction of environmental legislation.

7. For the general problems connected with estimation of production and cost functions for American cities see Emerson (1975).

8. The refuse created by inflowing commuters is not included because it figures under the heading of industrial waste.

9. A concise study proposing a way to measure noise, esthetics, health, safety, and general public satisfaction connected with waste collection is Blair, Hatry, and Don Vito (1970); an application is given in Blair and Schwartz (1972: chap. 3 especially).

10. See Schweizerischer Staedteverband, *Statistik der Schweizer Staedte* (various years) and Eidgenoessisches Statistisches Amt (1972).

11. *Correlation matrix:*

	U_1	U_2	U_3	Q_2	T_1	T_2	T_3	T_4	T_5	D_1	D_2
U_1	1.00	0.10	0.47	−0.01	−0.14	−0.10	−0.17	0.28	−0.17	0.08	0.10
U_2		1.00	0.29	−0.46	−0.14	0.52	0.37	−0.05	0.17	−0.07	−0.15
U_3			1.00	0.03	−0.02	0.05	−0.05	0.49	0.12	−0.04	0.13
Q_2				1.00	0.20	−0.44	−0.29	0.13	−0.03	0.16	0.32
T_1					1.00	−0.04	0.51	0.05	0.03	0.07	0.25
T_2						1.00	0.46	−0.02	0.06	−0.18	−0.19
T_3							1.00	−0.10	0.08	−0.06	−0.04
T_4								1.00	0.02	−0.05	0.14
T_5									1.00	0.01	−0.10
D_1										1.00	−0.38
D_2											1.00

12. The ceteris paribus condition is particularly important here because private enterprises usually have a positive profit and accordingly pay profit taxes. In the case in which public production seems advantageous to the taxpayers it should be noted that profit tax receipts fall when public production takes place. This may in the extreme lead to the effect that public production is more costly.

13. However, almost all "efficiency" studies exclusively compare costs with prices. As far as we are aware Savas (1976) is the only study differentiating between those two questions and concepts, but the implications are not fully worked out.

14. This markup is only approximate because insufficient information was available to calculate a representative average.

15. Similar results have been reached for American and Canadian cities (see Savas, 1976, and Kitchen, 1976).

16. No positive economies of scale in production could be found for the St. Louis City-County area (Hirsch, 1965:91) or for the localities of the Hartford and New Haven districts (Quigley and Kemper, 1976, chaps. 2 and 3). The only exception is Downing (1975:15), who found positive economies of scale for the 64 collection routes of the city of Riverside.

17. For a long time, economists and political scientists have disregarded the advantages of such a separation; but see V. Ostrom, Tiebout, and Warren (1961) for an exception.

18. As Young (1974:57) stresses: "While the execution of a performance bond (in the event of a contractor problem) can be a cumbersome matter, the bond is a valuable device, since financial institutions will presumably be unwilling to underwrite irresponsible firms."

REFERENCES

AHLBRANDT, R.S. (1973). "Efficiency in the provision of fire services." Public Choice, 16(fall):1-15.

ALCHIAN, A.A. (1967). Pricing and society. London: Institute of Economic Affairs.

ARROW, K.J. (1973). "The theory of discrimination." Pp. 3-33 in O.C. Ashenfelter and A. Rees (eds.), Discrimination in labor markets. Princeton, N.J.: Princeton University Press.

BAUMOL, W.J., and OATES, W.E. (1975). The theory of environmental policy. Englewood Cliffs, N.J.: Prentice-Hall.

BECKER, G.S. (1971). The economics of discrimination. Chicago: University of Chicago Press.

BISH, R.L., and WARREN, R. (1972). "Scale and monopoly problems in urban government." Urban Affairs Quarterly, 8(September):97-122.

BLAIR, L.H., HATRY, H.P., and DON VITO, P. (1970). Measuring the effectiveness of local government services: Solid waste collection. Springfield, Mass.: National Technical Information Service.

BLAIR, L.H., and SCHWARTZ, A.I. (1972). How clean is our city? A guide for measuring the effectiveness of solid waste collection activities. Washington, D.C.: Urban Institute.

DOWNING, P.H. (1975). "Intra-municipal variations in the cost of residential refuse removal." Unpublished manuscript, Virginia Polytechnic Institute and State University.

Eidgenoessisches Statistisches Amt (1972). Eidgenoessische volkszaehlung 1970, vol. 2: Gemeinden. Bern: Eidgenoessische Material- und Drucksachen Zentrale.

EMERSON, R.D. (1975). "The production and cost of police services." Unpublished manuscript, University of Florida.

FURUBOTN, E.G., and PEJOVICH, S. (1972). "Property rights and economic theory: A survey of recent literature." Journal of Economic Literature, 10(December):1137-1162.

GUERON, J.M. (1972). "Economics of solid waste handling and government intervention." Pp. 173-215 in S.J. Mushkin (ed.), Public prices for public products. Washington, D.C.: Urban Institute.

HIRSCH, W.Z. (1965). "Cost functions of an urban government service: Refuse collection." Review of Economics and Statistics, 47(February):87-92.

——— (1973). Urban economic analysis. New York: McGraw-Hill.

HIRSCHMAN, A.O. (1970). Exit, voice, and loyalty. Cambridge, Mass.: Harvard University Press.

KITCHEN, H.M. (1976). "A statistical estimation of an operating cost function for municipal refuse collection." Public Finance Quarterly, 4(January):56-74.

McKEAN, R.N., and BROWNING, J.M. (1975). "Externalities from government and non-profit sectors." Canadian Journal of Economics, 8(November):576-590.

OSTROM, E. (1973). "On the meaning and measurement of output and efficiency in the provision of urban police services." Journal of Criminal Justice, 1(summer):93-112.

OSTROM, V., TIEBOUT, C.M., and WARREN, R. (1961). "The organization of government in metropolitan areas: A theoretical inquiry." American Political Science Review, 55(December):831-842.

PIER, W.J., VERNON, R.B., and WICKS, J.H. (1974). "An empirical comparison of government and private production efficiency." National Tax Journal, 27(December): 653-656.

QUIGLEY, J.M., and KEMPER, P. (1976). The economics of refuse collection. Cambridge, Mass.: Ballinger.

SAVAS, E.S. (1976). "Evaluating the organization of service delivery: Solid waste collection and disposal." Unpublished manuscript, Columbia University.

Schweizerischer Staedteverband (ed., annual). Statistik der Schweizer Staedte. Zürich: Schweizerischer Staedteverband.

STIGLER, G.J. (1971). "The theory of economic regulation." Bell Journal of Economics and Management Science, 2(spring):3-21.

U.S. Department of Health, Education, and Welfare (1970). Policies for solid waste management. Washington, D.C.: U.S. Government Printing Office.

YOUNG, D.R. (1974). "The economic organization of refuse collection." Public Finance Quarterly, 2(January):43-72.

9

The Service Paradox:
Citizen Assessment of Urban Services in
36 Swedish Communes

BENGT OWE BIRGERSSON

☐ IN SWEDEN, local government is often justified as facilitating the adaptation of public services to local variations in need. Where services are provided by local governments, it is possible to vary the quality and quantity of such services in different localities. The shorter distance between elected decision makers and citizens also makes it easier for citizens to communicate their wishes and demands to local decision makers. According to this view, the smaller the commune (the unit of local government in Sweden) the easier it is to adapt the standard of services to citizens' desires.

This line of reasoning conflicts with the view that small communes lack sufficient resources—in terms of revenue and personnel—to produce services of the scope and quality demanded by citizens. For this reason it has been judged necessary to merge smaller communes into larger units. In the process, the number of communes has decreased since 1952 from 2,500 to 278.

The nature of local democracy changes substantially when communes become larger. The evolution of local government in Sweden can be characterized as a transition from a local government with features of direct participatory democracy to a modern, functionally organized, representative system. In this article, we explore the relationships between commune size, citizen satisfaction, and the standard of local government service. Two questions are posed:

- Do larger communes generally offer a "better" standard of local government service than the smaller communes?
- Are the citizens more satisfied with local government service in the communes where service levels are highest?

These two questions served as the focal point of a study conducted with the Local Government Research Group at the University of Stockholm. The study in its entirety is presented in a Swedish doctoral dissertation (Birgersson, 1975). This article briefly summarizes the study and some of its main results.

DESIGN AND METHODS

THE SAMPLE OF COMMUNES

Most of the Local Government Research Group's studies have been conducted in a sample of 36 Swedish communes (Westerstahl, 1971). In this sample the communes have been stratified into six different categories based on size and density of population, as presented in Table 1. In each cell, three of the communes have a nonsocialist majority in the local government council. In the other communes in the cell, the Social Democrats and Communists together enjoy a majority. Due to changes in apportionment of local government units during the course of the research, it was necessary to exclude two communes in cell IA and one commune in cell IIA from the study presented in this article. Thus, the study comprised a total of 33 communes.

TABLE 1

THE LOCAL GOVERNMENT RESEARCH GROUP'S SAMPLE OF COMMUNES

Population Density[a]	I Less than 8,000	II 8,000 to 30,000	III More than 30,000
A. More than 90%	6	6	6
B. 30 to 90%	6	6	
C. Less than 90%	6		

a. Percentage of inhabitants living in built-up areas.

THE STANDARD OF LOCAL GOVERNMENT SERVICE

Accurate measurement of the standard or level of local govern-ment services entails sizable difficulties. Not only is "service standard" a multidimensional concept but it probably means different things to different people. It must also be viewed in relation to variations in need. In public policy studies, expenditures in various areas by local governments or individual states in the United States are used as indicators of the importance that decision makers attach to various sectors or as a standard measurement of service output (see, e.g., Dye, 1972; Scharkansky, 1967).

This approach, however, is not meaningful in the case of Swedish communes. Various studies of local government activity have revealed vast differences in the productivity of different communes. Thus it would be completely meaningless to use a measure based on expenditures per inhabitant as an indicator of the service standard.

The method of measuring local government services used in this study is not without difficulties either (Birgersson, 1975, chap. 4). But within the limits of a rational use of resources, the method appears, at present, to be the more plausible one.

Local government activities were divided into 12 sectors:

- Town planning
- Housing
- Employment opportunities
- Streets, roads, and traffic
- Waterworks and sewage
- Schools
- Day nurseries
- Home help for families with children and the elderly
- Old-age homes
- Social assistance
- Leisure activities
- Libraries and other cultural activities

In each sector several different measures of performance were selected, and, to the extent possible, an effort was made to determine service needs (in terms of the size of clientele groups, etc.). These measures of performance and need were analyzed by service area. Using these analyses as a guide, 12 operationalizations within 9 areas were chosen to form a single index.

Area of Service	Operationalization
1. Town planning	1. Proportion of housing construction in 1966-1969 located on land owned by the commune
2. Housing	2a. Proportion of housing constructed by public utilities
	2b. "Surplus" of dwellings in relation to number of households
3. Streets and roads	3. Proportion of asphalted streets in built-up areas
4. Home help	4a. Proportion of citizens older than 70 who had help (from commune) in the home
	4b. Proportion of families with children who had help (from commune) in the home
5. Old-age home	5. Number of places in old-age homes per 100 citizens older than 70
6. Day nurseries	6a. Number of places in day nurseries per 100 children under 5 years of age
	6b. Number of children registered in play schools per 100 six-year-olds
7. Social assistance	7. Amount of social assistance (expenditure per inhabitant)
8. Leisure activities	8. Budget expenditure for sports and leisure-time activities (expenditure per inhabitant)
9. Libraries	9. Library expenditures (expenditure per inhabitant)

Three areas are not represented in the index. In two of the areas, "schools" and "waterworks," service levels varied little among communes. In the case of "employment opportunities," the communes have limited possibilities to act. Responsibility for labor market policy in Sweden is primarily a task of the national government. Among the remaining services we found considerable variation among different communes: variations which are a result of policies in the communes.

A major problem in combining the various service measures into one comprehensive measure was the widely differing character of the measurements. What value is to be assigned to the proportion of housing constructed by public utility companies in relation to the amount of social assistance measured in expenditures per inhabitant? In view of these difficulties, the communes were simply ranked in each area of service on the basis of the selected measurements of service. Thus for each service area a score was obtained indicating the level of local government service in a particular commune in relation to the corresponding service in the rest of the communes. By subsequently computing the average rank for all the service areas, a measure was obtained showing how extensive the commune's service is on an average in comparison with the other communes.

CITIZEN PERCEPTIONS OF SERVICE

Data on citizen attitudes toward local government services were collected through an interview survey in connection with the 1970 general election.[1] The survey included all 36 communes in the sample. The sample of interviewed persons consisted of 65 people in each commune. A total of 91% of the planned interviews were actually conducted.

For each of the 12 service areas the interviewee was asked a number of questions concerning his or her attitude toward the particular service. Parallel questions were asked for each service area. As an illustration, the question concerning housing construction had the following wording:

A. First, concerning the availability of housing, do you think that on the whole this issue has been well handled here in [the name of the commune], or are you dissatisfied with some aspect?

The alternative answers were:

(1) On the whole handled well

(2) Dissatisfied with some aspect

(3) No opinion

In appropriate cases three follow-up questions were asked:

B. What are you dissatisfied with?

C. Is this due to yourself or any of your immediate family?

D. Do you know of any *other* people here in [the name of the commune] who are dissatisfied with this aspect?

The purpose of questions B, C, and D was primarily to avoid as far as possible service demands which were first formulated during the interview.

A comprehensive measure of the citizens' assessment of services was devised by summing the answers to the A questions for each respondent. The measure obtained in this way—the number of areas in which the interviewed person expressed dissatisfaction with the service—was dichotomized by placing persons who were dissatisfied in a minimum of six areas in a group with a "wide range of demands" while the remaining persons were designated as having a "narrow range of demands." The validity of this measure of service assessment was checked by comparing it with a question concerning the interviewee's evaluation of local government service in general.[2]

In Figure 1 a correlation appears between the range of demands and the overall evaluation of local government service. Among the respondents who were completely satisfied or on the whole satisfied, only 6% were dissatisfied with some aspect in more than five areas of service. Among those who were mainly or completely dissatisfied with the service, more than half of the group exhibited a large range of demands. It can be further observed in the figure that persons who felt unable to make an overall judgment on local government service in general also had a narrow range on demands. In view of these results the measure of the "range of demands" was considered a useful indicator of the citizens' general assessment of service.

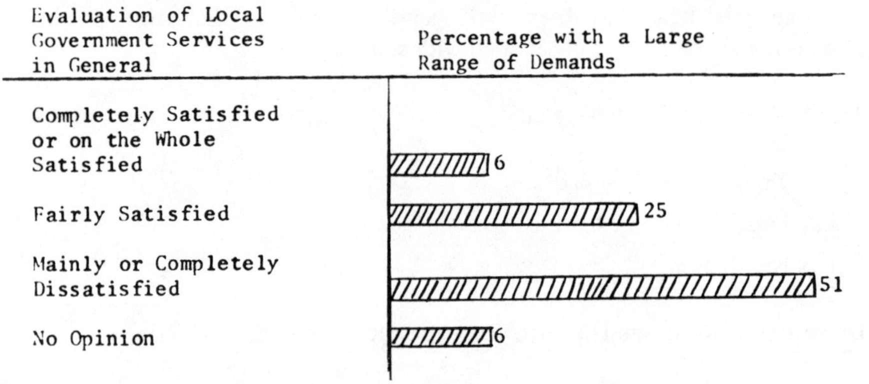

Evaluation of Local Government Services in General	Percentage with a Large Range of Demands
Completely Satisfied or on the Whole Satisfied	//////////// 6
Fairly Satisfied	///////////////////////// 25
Mainly or Completely Dissatisfied	/// 51
No Opinion	//////////// 6

Figure 1: EVALUATION OF LOCAL GOVERNMENT SERVICE IN GENERAL

TABLE 2

SERVICE INDICES FOR VARIOUS TYPES OF COMMUNES

	Arithmetic Mean Size of Population			Median Size of Population		
	I	II	III	I	II	III
Population Density[a]	Less than 8,000	8,000 to 30,000	More than 30,000	Less than 8,000	8,000 to 30,000	More than 30,000
A. More than 90%	18.5	14.3	12.1	18.0	12.9	12.2
B. 30 to 90%	16.6	16.1		16.5	15.7	
C. Less than 90%	20.3			20.1		

a. Percentage of inhabitants living in built-up areas.

THE STANDARD OF SERVICE AND SERVICE ATTITUDES

What relationship exists between the standard of service and service attitudes? Table 2 presents variations in the service index by type of communes. Since the index is based on rank ordering, *a low value is a sign of high standard of service on the average and vice versa.* The table clearly shows higher standards of service in the larger communes as measured by service indicators. With the exception of the small densely populated communes in category IA, a correlation with the degree of population density can be noted. In other words, communes having a major portion of the population living in built-up areas also have higher levels of local government service.

Table 3 presents the proportion of interviewed persons who declared their dissatisfaction with some aspect of service in a minimum of 6 out of the 12 service areas. The table discloses that dissatisfaction with local government service is most widespread in

TABLE 3

PERCENTAGE OF INTERVIEWED PERSONS WITH A LARGE RANGE OF DEMANDS IN THE 1970 INTERVIEW SURVEY: BY TYPE OF COMMUNE

Population Density[a]	I Less than 8,000	II 8,000 to 30,000	III More than 30,000
A. More than 90%	10	17	26
B. 30 to 90%	12	17	
C. Less than 90%	9		

a. Percentage of inhabitants living in built-up areas.

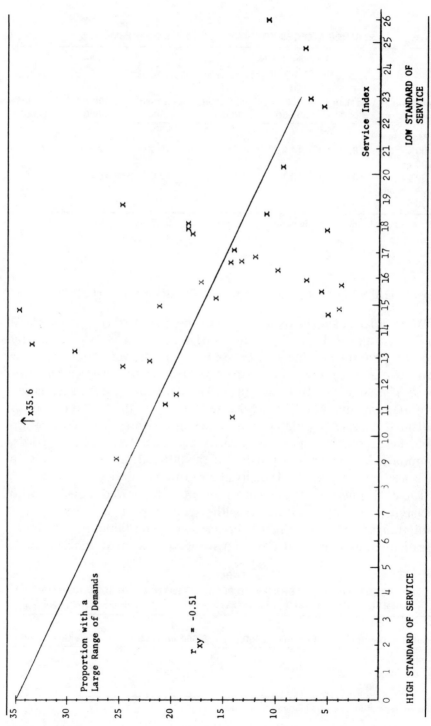

Figure 2: CORRELATION BETWEEN THE SERVICE INDEX AND THE PROPORTION OF RESPONDENTS WITH A LARGE RANGE OF DEMANDS

the large towns, where 26% of the respondents revealed a wide range of demands. Next come the medium-sized communes with 17%, and last the small communes where respondents were most satisfied in both sparsely and densely populated communes. In these two types of communes only 9% and 10% respectively had a wide range of demands.

A comparison of Tables 2 and 3 indicates that the proportion of respondents with a wide range of demands was highest in the communes where the service standards are highest. A comparison of the standard of service and service attitudes in each particular service area verified this impression. The result is summarized in Figure 2, where the correlation has been tested using regression analysis. The overall correlation between the proportion of respondents with a wide range of demands and the service index is clearly negative (r_{xy} = −0.51). This means that the lower the average service rank a commune has (i.e., the higher the standard of service), the larger the proportion of interviewed persons who declared their dissatisfaction with some aspect of service within at least 6 of the 12 service areas. Citizens appear to be more satisfied in the communes where services are lower.

This finding, which we have called the "service paradox," was subjected to a series of analyses to determine the extent to which it could be a result of the research techniques used in the study or a result of factors having no direct connection with local government service. Viewed from the perspective of the individual, it is likely that the assessment of service depends not only upon commune-related attributes such as the *standard of local government service* and *the environmental features of the commune* but also upon *individual attributes* influencing service needs and upon *general political views about the nature of the community's role in the individual's welfare.* It can also be assumed that the degree of involvement in politics influences the assessment of the results of local government activities.

The subsequent study of the "service paradox" supports the following conclusions:

The citizens' assessment of service varied among different groups. For example, respondents who are active, well-educated, and interested in politics more frequently revealed a wide range of demands than the other respondents.

Some categories of citizens with especially demanding attitudes regarding a commune's provision of services constituted a larger proportion of the citizens in communes where services standards are higher (i.e., larger communes). At the same time, however, other categories of citizens with especially demanding attitude comprised a smaller proportion in the types of communes where the service standard was lowest.

Particularly the second conclusion suggests that the service paradox may depend upon how individuals with differing preferences are distributed among the various types of communes. In view of the importance of personal attributes in shaping one's basic attitudes toward services, we conducted a multivariate analysis of the service standard, personal attributes, type of commune, and service attitudes. This analysis is presented in the next section.

MULTIVARIATE ANALYSIS OF THE SERVICE STANDARD, PERSONAL ATTRIBUTES, TYPE OF COMMUNE, AND SERVICE STANDARD

In this section, we examine the relationship between range of demands and service standard controlling for type of commune (i.e., the population density/size classification of the commune) and for personal attributes.

As indicated above, there is a relatively strong *negative* relationship (r_{xy} = -0.51) between the service rank of a commune and the proportion of persons with a wide range of demands: the lower the rank of the commune (i.e., the higher the standard of service), the larger the proportion of persons indicating dissatisfaction with at least 6 of 12 service areas. As one can see from Figure 2, however, there also appear to be two clusters of communes where the correlation between rank of commune and range of demands is *positive* rather than negative. Within each of these clusters, a higher proportion of persons are dissatisfied in communes with lower service levels than in communes with higher service levels.

Each of these clusters represents different population density/size classifications of communes. Because there is a positive relationship between the standard of service and commune type (the larger and more urbanized the commune, the higher the level of service), it is possible that the "service paradox" reflects, in part, a generally higher level of dissatisfaction among citizens living in the larger and more urbanized communes. This generally higher level of dissatisfaction may, in turn, reflect differences in needs and/or citizen demands for services in the larger communes.

In constructing the service index, we chose as far as possible, items where the amount of service was examined in relation to the size of clientele groups. It is possible that this control for variations in "need" is inadequate—that the service need (or level of demand) is proportionately greater in urban environments. Thus, while service levels measured in relation to population may be higher in larger communes, they may be lower in relation to citizen needs and/or demands. One way to examine this hypothesis is to group the communes in each size/density category according to standard of service and then analyze citizen attitudes in each subcategory. If it is a correct assumption that the level of service demand is caused by urbanization *and* if it also can be assumed that citizen dissatisfaction reflects in part a failure to provide the desired level of services, the proportion of respondents displaying a wide range of demands ought to be greatest in each size/population density category where the standard of service is low. Moreover, the proportion of persons with a large range of demands should also be greatest in the larger, more urbanized communes—regardless of the level of services provided.

In Table 4, the communes in each size/population density category were rank-ordered according to their service index score. The three communes which had the lowest standard of service were placed in the group with low service; the remaining three were assigned to the group with a high standard. The table presents the average and median proportion of respondents displaying a large range of demands in the 12 size/population density/service categories.

Unfortunately, a clear-cut picture does not emerge from the data. In Table 4, the *average* proportion with a large range of demands is highest in the low standard communes in only two of the six size/population density classes. In one class (IB), the difference is minute, but in three classes of communes (IC, IIA, IIB), the proportion of residents with a large range of demands is actually highest in communes which have high service levels.

The medians offer a picture which is more in accord with the assumption concerning the effect of urbanization. In all the size/population density classes except IC, the proportion with a broad range of demands is largest in the three communes where the standard is lowest. The result can be interpreted as support for the

TABLE 4

PERCENTAGE OF RESPONDENTS WITH A LARGE RANGE OF DEMANDS:
BY LEVEL OF SERVICE PROVIDED BY COMMUNE AND
SIZE/POPULATION DENSITY OF COMMUNE

	Arithmetic Mean Size of Population					Median Size of Population					
	Less than 8,000		8,000 to 30,000		More than 30,000		Less than 8,000		8,000 to 30,000		More than 30,000
	Standard of Service		Standard of Service		Standard of Service		Standard of Service		Standard of Service		Standard of Service
Population Density[a]	Low	High	Low	High	Low	High	Low	High	Low	High	Low	High
A. More than 90%	10.6	9.3	13.1	23.8	29.5	21.5	10.6	9.3	23.8	21.7	34.6	20.3
B. 30 to 90%	11.9	12.0	16.5	17.9			13.7	9.5	18.0	15.7		
C. Less than 90%	7.3	10.8					6.3	11.5				

a. Percentage of persons living in built-up areas.

hypothesis that the negative correlation between the level of service and citizen satisfaction is partially an effect of the environment leading to greater demands in larger towns than in smaller and more sparsely populated communes. Against the background of the averages in Table 4, it is however doubtful *how much importance* should be attached to this interpretation.

Are there remaining variations in the composition of the population of individuals which can influence and "disturb" a correlation which would emerge more clearly if it were possible to control for these variations? In order to shed light on this question, the analysis was carried one step further. For this purpose we used a technique of analysis which is now well-known—the so-called tree analysis (automatic interaction detector analysis, or AID analysis). This technique, which resembles multiple regression but makes fewer demands of the scale level of the independent variables, was developed at the Institute for Social Research, Ann Arbor, Michigan. It has been described in numerous publications (Sonquist and Morgan, 1965; Sonquist, 1969; Brantgärde, 1971).

The tree analysis can be described as a technique to achieve the most effective classification possible of a number of independent variables with respect to their ability to predict the value of a dependent variable. First the entire sample is treated as a group, and the computer program goes interactively through all conceivable classifications of the sample with regard to all independent variables in order to divide the sample into two groups so that the proportion of explained variance in the dependent variable is maximized. The process then continues in the same way for each of the created subgroups, which in turn are subsequently divided into several smaller groups so that the amount of explained variance is maximized. The dividing process continues until no further divisions are possible. In this connection it is possible for the analyst, by setting up specific criteria, to determine how long the process should continue. In this study, one criterion was that no group was to consist of fewer than 25 individuals. Thus a group consisting of fewer than 50 individuals could not be further divided. An additional criterion was that the reduction of "unexplained" variance should be at a minimum 0.6%.

The independent variables in an AID analysis can be of a nominal scale level or higher. The dependent variable, on the other hand, must be either a metric scale or consist of a dichotomy. As a measure of service assessment we have so far used the variable "range of demands" dichotomized into two categories—wide and narrow. In a tree analysis, difficulties arise if the dependent variable consists of a dichotomy in which one of the categories is substantially larger than the other. Therefore, in the analysis described below, the range of demands is measured as *the number of service areas in which each interviewed person declared dissatisfaction with some aspect. Hence the dependent variable is composed of an interval scale with values from 0 to 12.*

Here, using a tree analysis, we shall allow individual attributes to operate as long as possible—i.e., explain as large a proportion of the variance in the dependent variable as possible. Such an analysis results in a number of final groups in which the populations of individuals are as homogeneous as possible with respect to their assessment of services. By subsequently examining once again the correlation between the service index and the type of commune and the proportion of dissatisfied respondents but only among individuals from the relevant final group from the AID analysis, any possible correlations between the aforementioned variables ought to appear as far as possible "undisturbed" by the influence of individual attributes.

Since the purpose of the tree analysis is only to control for "disturbing" effects of the individual-related variables, the relationship among these different variables is of secondary interest in this context. In view of this, the AID run was carried out in one step with

12 individual-related variables—which in previous analysis revealed covariance with the assessment of services. These 12 variables were:

- Frequency of discussing local government issues with friends
- Residential area's character
- Age
- Knowledge of politics (number of correct answers concerning the jurisdiction of local government)
- Involvement in efforts to change communal conditions that were regarded as unsatisfactory
- Interest in politics
- Voting for a party in the local election
- Party membership
- Membership in organizations
- Education
- Acquaintance with an elected official
- Marital status

The results can be observed in Figure 3. As can be seen from the "tree" in the figure, the average individual is dissatisfied in 3.3 areas of the service. The variable exhibiting the greatest covariance with dissatisfaction is "discussing local issues among friends." When persons who did and did not often discuss local issues are divided into two groups, the variance is reduced by roughly 7%. The data in Table 5 reveal that age comes next (the variance had then been reduced by nearly 6%—BSS/TSS = 0.059). Then follows efforts to change conditions and education.

From the tree it can be further observed that dissatisfaction is greatest among the group who often discuss local issues and who are 59 or less in age. These 383 persons are, on an average, dissatisfied with 4.6 areas of service. The group displaying the lowest stated dissatisfaction is characterized by never discussing local issues with anyone in their surrounding and by being over 60 years of age. This group comprises 253 persons in the sample, and on an average they are dissatisfied with 1.8 areas of service.

The tree analysis can produce misleading results in instances of covariation between independent variables. In such cases the division will occur on the basis of the independent variable which statistically covaries best with the dependent variable irrespective of whether the

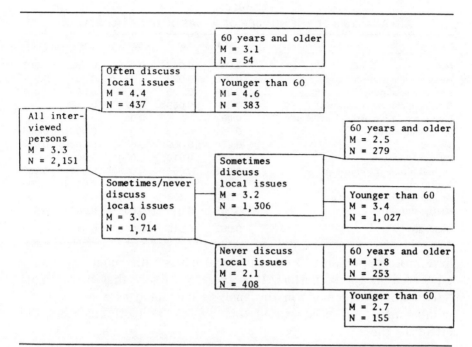

Figure 3: TREE ANALYSIS WITH INDIVIDUAL-RELATED VARIABLES AS INDE-
PENDENT VARIABLES AND THE NUMBER OF AREAS OF DECLARED
DISSATISFACTION AS THE DEPENDENT VARIABLE

other independent variable is theoretically the most interesting. The effect of the latter variable is neutralized, and it is very probable that it never will be capable of asserting itself in the tree emerging as a result of the analysis. Using the tree analysis technique without paying attention to this possibility can lead to totally erroneous results.

In Figure 3 it is evident that only 2 of the 12 variables "participated" in the analysis. At the same time we know from our other analyses that most of the variables reveal a clear covariation with the range of demands. The effect of the 10 variables which did not participate in the tree has thus been blocked by the 2 variables which contributed to the groupings in the figure.

On reflection it seems reasonable to regard the variable "discussion frequency" as dependent on several other independent variables. Persons who are interested in politics can be expected to discuss local government issues more frequently than other individuals. Party members and individuals personally acquainted with some local official as well as probably persons living in urbanized communities

TABLE 5
AID ANALYSIS OF THE NUMBER OF AREAS OF DISSATISFACTION

Variables	Set 1	Set 2	Set 3	Set 4	Set 5
Age	0.059	**0.043**	0.051	**0.033**	**0.055**
Residential areas character	0.018	0.032	0.011	0.007	0.019
Knowledge of politics	0.029	0.000	0.029	0.0008	0.051
Efforts to change conditions	0.040	0.009	0.024	0.016	—
Frequency in discussing local issues	**0.067**	—	0.053	—	—
Interest in politics	0.031	0.003	0.031	0.003	0.000
Voting	0.004	0.009	0.003	0.008	0.027
Party membership	0.000	0.019	0.000	0.000	—
Organizational membership	0.017	0.006	0.008	0.004	0.003
Education	0.034	0.021	0.018	0.012	0.002
Acquainted with elected official	0.000	0.014	0.000	0.004	0.001
Marital status	0.003	0.006	0.006	0.003	0.013
BSS/TSS$_T$	0.067	0.010	0.036	0.017	0.008

have more opportunity to discuss local issues with somebody who is interested in these matters. It is therefore likely that the variable "discussion frequency" in the analysis has acted as a proxy for a number of other variables and thereby has indirectly absorbed the effects of these.

The combined explanatory capacity of the individual-related variables is limited. As brought out in Table 5, the tree analysis in Figure 3 only succeeded in reducing the variance in the dependent variable by approximately 14%. The question, however, is whether or not the distribution of individuals among different types of communes is skewed to such an extent that it can influence the result of the correlation analysis of service attitudes and the type of commune.

The tree analysis in Figure 3 resulted in six final sets. They were:

(1) Persons who often discuss local issues and are 60 or more years of age. This group consists of 54 persons who are dissatisfied with an average of *3.1 areas.*

(2) Persons who often discuss local issues and are 59 or less years of age. This group consists of 383 persons who are dissatisfied with an average of *4.6 areas.*

(3) Persons who sometimes discuss local issues and are 60 or more. This group consists of 279 persons who are dissatisfied with an average of *2.5 areas.*

(4) Persons who sometimes discuss local issues and are 59 or less years of age. This group consists of 1,027 persons who are dissatisfied with an average of *3.4 areas.*

(5) Persons who never discuss local issues and are 60 or more years of age. In this group there are 253 persons, and they are dissatisfied with an average of *1.8 areas*.

(6) Persons who never discuss local issues and are 59 or less years of age. In this group there are 155 persons who are dissatisfied with an average of *2.7 areas*.

In Figure 4 the interviewed persons in the six final sets have been grouped according to the type of commune in which they live. The

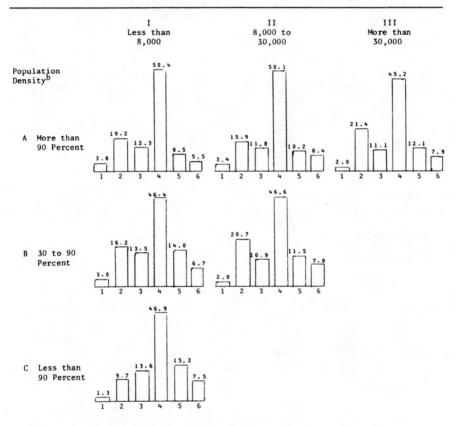

^a1. 60 or more years of age and often discuss local issues
2. Younger than 60 and often discuss local issues
3. 60 or more years of age and sometimes discuss local issues
4. Younger than 60 sometimes discuss local issues
5. 60 or more years of age and never discuss local issues
6. Younger than 60 never discuss local issues

^bPercentage of persons living in built-up areas

Figure 4: PROPORTION OF INTERVIEWED PERSONS BELONGING TO DIFFERENT FINAL GROUPS IN THE AID ANALYSIS: BY TYPE OF COMMUNE^a

TABLE 6

PERCENTAGE OF RESPONDENTS WITH A LARGE RANGE OF DEMANDS:
BY TYPE OF COMMUNE AND FINAL GROUPS FROM THE TREE ANALYSIS

	Size of Population		
Population Density[a]	I Less than 8,000	II 8,000 to 30,000	III More than 30,000
a. Final Set 1			
A. More than 90%	0	16.6	22.2
(N)	(9)	(10)	(7)
B. 30 to 90%	25.0	0	—
(N)	(11)	(8)	—
C. Less than 30%	0	—	—
(N)	(5)	—	—
	N Total = 11.1		
	N Total = 54		
b. Final Set 2			
A. More than 90%	22.7	31.2	49.1
(N)	(45)	(47)	(73)
B. 30 to 90%	30.0	32.4	—
(N)	(59)	(74)	—
C. Less than 30%	28.9	—	—
(N)	(35)	—	—
	N Total = 34.5		
	N Total = 383		
c. Final Set 3			
A. More than 90%	0	5.0	11.2
(N)	(29)	(35)	(38)
B. 30 to 90%	6.8	20.0	—
(N)	(49)	(39)	—
C. Less than 30%	3.2	—	—
(N)	(61)	—	—
	N Total = 7.5		
	N Total = 279		
d. Final Set 4			
A. More than 90%	9.2	20.1	24.1
(N)	(118)	(148)	(154)
B. 30 to 90%	9.7	16.7	—
(N)	(169)	(166)	—
C. Less than 30%	7.7	—	—
(N)	(177)	—	—
	N Total = 14.0		
	N Total = 1,027		
e. Final Set 5			
A. More than 90%	3.1	1.8	13.1
(N)	(20)	(30)	(41)
B. 30 to 90%	3.4	1.2	—
(N)	(51)	(41)	—
C. Less than 30%	1.9	1.2	—
(N)	(55)	—	—
	N Total = 3.6		
	N Total = 253		

TABLE 6 (continued)

Population Density	Size of Population		
	I Less than 8,000	II 8,000 to 30,000	III More than 30,000
f. *Final Set 6*			
A. More than 90%	4.2	14.8	14.3
(N)	(13)	(25)	(27)
B. 30 to 90%	7.5	0	—
(N)	(25)	(28)	—
C. Less than 30%	22.9	—	—
(N)	(27)	—	—
	N Total = 10.3		
	N Total = 155		

a. Percentage of inhabitants living in built-up areas.

figure initially gives the impression that variations in the pattern of distribution of individuals from the various final sets are very limited.

In the small, sparsely populated communes and the small communes of mixed character—IC and IB—the groups of least-demanding individuals (groups 3, 5, 6) are, however, the most common, and persons in the group of most dissatisfied individuals (Group 2) are relatively few (especially in IC). The lower proportion of persons with a wide range of demands in these communes can be partially explained by these communes having a larger proportion of citizens who are older and—probably—also less interested in and less informed about local government issues (at any rate they discuss local issues less frequently with their friends). Among the remaining types of communes, the individual attributes do not contribute much to explaining the difference in the proportion of persons with a wide range of demands. Against the background of the service standard being definitely higher in absolute terms in the large towns (the service index was 12.1 as compared to 18.5) the earlier observed service paradox seems to have gained additional support through the tree analysis.

When the analysis refocuses on the size/density of the commune in Table 6 and the correlation between type of commune and service attitudes is examined in each final set in the tree analysis, we are faced with the common problem in multivariate analysis of cross-tables—the small number of observations. This gives factors for which we have not controlled considerable latitude and leads to the analysis giving a "thicket-like" impression that is difficult to interpret. Table 6 presents the proportion of interviewed persons revealing a large range of demands in each final group (M_{TOTAL}) and the number of interviewed persons in each final group (N_{TOTAL}). The table further indicates the proportion with a large range of demands in each type of commune as well as the number of interviewed persons in each type of commune.

In final sets 1 and 6 the number of interviewed persons is so small that it would be meaningless to examine in detail the distribution of interviewees by type of commune. In the remaining final sets, however, the number of interviewed persons is larger. Therefore, what picture does Table 6, b-e provide of the relationship between the type of commune and service attitudes? If the differences between the various types of communes were caused by variations in the composition of the population of individuals, in each final group in Table 6 no major differences would remain between the various types of communes. However, in three of the final groups the proportion with a large range of demands is greatest in IIIA. In the fourth group the proportion is larger only in IIB. Nonetheless it is difficult visually to decipher a clear pattern from the table. One way of summarizing Table 6, b-e, is to compute for each type of commune the average deviation (M_{RES}) from the mean of each final set (M_{TOTAL}) according to the formula,

$$M_{RES} = \frac{\sum_{i=1}^{4} \{ (a_i) }{4}$$

where $a_i = (M_{TOTAL} - \bar{x})$

$\bar{x}$ = proportion with a large range of demands in the particular type of commune.

The results are presented in Table 7. As can be seen in the table, the proportion with a large range of demands on the average is 9.5%

TABLE 7

AVERAGE DEVIATION (M_{RES}) IN FINAL GROUPS 2-5 OF THE PERCENTAGE OF RESPONDENTS WITH A LARGE RANGE OF DEMANDS: BY TYPE OF COMMUNE

Population Density[a]	I Less than 8,000	II 8,000 to 30,000	III More than 30,000
A. More than 90%	−6.2	−0.4	+9.5
B. 30 to 90%	−2.4	+2.7	
C. Less than 90%	−4.5		

a. Percentage of inhabitants living in built-up areas.

higher than expected in commune type IIIA and 6.2% lower than expected in IA. The picture emerging in Table 7 is largely the reverse of Table 2, which presented the average service rank in the different types of communes. The service paradox appears to have received additional support by the tables presented above (Tables 6 and 7).

The analysis above has the advantage of being easily understood and simple to carry out. However, two criticisms can be leveled against it. First, more than 200 individuals (those in final sets 1 and 6 in the AID analysis) have been excluded. Second—and this is probably a more important objection—the 253 individuals in final set 5 have as much influence on the results of the analysis as the 1,027 individuals in final set 4. To the extent that there is an interaction between the final sets and the type of commune causing the pattern to vary among different final sets, this lack of weighting can mean that the result is misleading.

In order to eliminate these two weaknesses, we conducted an additional analysis in which responses of all interviewed persons are included with equal weighting. Taking the number of interviewed persons for each type of commune from each final set, we computed the expected proportion of interviewed persons with a large range of demands for each type of commune. Then the observed value for the proportion with a large range of demands was subtracted from the computed expected value.[4] The results are presented in Table 8.

On all essential points Table 8 provides the same picture as Table 7. However, cell IB has changed places with IA and IC. In relation to what is to be expected in view of the composition of the population of individuals, the proportion with a large range of demands is least in IB followed by IC, IA, IIB, and IIIA. The negative correlation between the size of the commune and service attitudes persists. A new comparison with Table 2 still provides a definite impression of a negative correlation with the standard of service. If the *types* of

TABLE 8

AVERAGE DEVIATIONS FROM THE EXPECTED PERCENTAGE OF RESPONDENTS WITH A LARGE RANGE OF DEMANDS: BY TYPE OF COMMUNE

Population Density[a]	I Less than 8,000	II 8,000 to 30,000	III More than 30,000
A. More than 90%	−4.9	−1.9	+10.0
B. 30 to 90%	−9.1	+1.0	
C. Less than 90%	−4.0		

a. Percentage of inhabitants living in built-up areas.

communes are rank-ordered according to the measurement of service standard in Table 2 b, and according to the proportion of respondents with a large range of demands before controlling for individual-related variables (Table 3) and after the control in this section (Table 8), Spearman's Rho can be employed to compute the correlation. This control leads to the correlation coefficient dropping from 0.88 to 0.77. Our service paradox has thus survived this final attempt to dispose of it. The proportion of respondents with a wide range of demands is greatest in the types of communes where service standards, according to all our measurements, are the highest.

CONCLUSION

The Local Government Research Group's study of services raises more questions than it answers. It demonstrates that service levels are higher in most respects in the larger communes. Thus the answer to our first question was "yes." In contrast, the answer to the second question was a definite "no." Citizens were not more satisfied with the service in the communes supplying higher levels of service. This finding, which we regarded as surprising against the backdrop of the debate on local government services, survived all the controls which we carried out (all of which are not presented here). Even taking into account the methodological weaknesses of our measurements of the standard or level of service, we must therefore accept the existence of the so-called service paradox until there is evidence to the contrary.

What, then, is the explanation for this service paradox? One explanation might be that urbanization is associated with increased needs for communal services—needs that the larger and more densely populated communes are, by and large, unable to meet. This problem is probably greatest in countries like Sweden where urbanization during the postwar period has been very rapid. It is in the larger, more urbanized communes that we find both the highest service levels and the largest proportion of persons who are dissatisfied. The smaller communes on the other hand, have chosen to provide lower levels of service (at lower tax rates) and seem to be able to do so without substantial dissatisfaction on the part of persons living in these communes.

Part of the explanation for the service paradox may, then, be that urbanization is associated with increased *needs* for services. It is also possible, however, that the results mirror a more general problem of rising expectations. A tempting thought is to suggest that an expanded level of services whets the appetite. When a commune does not provide a particular kind of service, the lack is felt by comparatively few persons. When the commune (and other communes of the same size and density) expands its activity, an increasing number of people become aware of the service and wish to utilize it. The increase in demand tends then to grow more rapidly than service capacity, and during a transitional period there paradoxically arises a mounting dissatisfaction which ultimately is caused by a rise in the standard. If this is actually the case, the continuous rise in the standard of public services mainly following the urbanization process may also mean that an ever growing dissatisfaction is unconsciously built into society.

An interesting question is, of course, how the innovation process starts—how new "service ideas" spread from commune to commune. In an article in *Scandinavian Political Studies* (1975), Daniel Tarschys points to the producer perspective as an explanation of the expansion of the public sector in Sweden and several other Western countries. If this perspective is applied to the local government sector, the innovation process could be sketched in the following way. The "specialists," i.e., professionals, in a particular administrative area are interested in an expansion of service or in a new program. Since professionals are most common in the larger communes where the administration is the most specialized, the new ideas are initially tried out in larger communes. This means that the innovation process first begins in the larger communes. Dissatisfaction with the inadequate dimensions of the new service emerges in the larger urbanized communes long before citizens in other communes have even become acquainted with the service. The professional administrators in specialized fields exchange ideas and experiences, for example, at conferences and through articles in professional journals. The Association of Swedish Communes organizes an extensive program of courses through which new ideas are disseminated to the communes. Through these various channels the new ideas are thus spread to the communes which have not yet accepted them. It is likely that "opposition" is weakest in the somewhat larger communes where professional administrators are a more dominant element and there is greater budgeting flexibility. In view of these circumstances the larger communes accept the innovations before the small communes, whose administration is more dominated by elected officials.

Our study does not provide any empirical data supporting these speculations. Rather they ought to be viewed as suggestions for additional research concerning expansion of the local government sector. The interesting point about this hypothetical reasoning is that it can serve as an alternative to the common conception that innovations in public service delivery are initiated through the wishes and demands of the citizens. Here the initiative originates from the employees and by what at least partially can be conceived of as "the self-interest of bureaucrats" (Downs, 1967). Perhaps the expansion of the public sector in several countries is propelled by a kind of planned demand on the part of the bureaucracy.

NOTES

1. Lennart Mansson at the University of Gothenburg was responsible for the construction of the questionnaire used in the survey. The coding, punching, and program tests were done at the University of Gothenburg. The analysis was done by the author at the Stockholm computer center.

2. The question read: "We have talked about the local government's efforts in various service areas. How would you in general evaluate the local government service here in [the name of the commune]?" The range of answers were:

(1) I am completely satisfied with local government service.

(2) I am on the whole satisfied with local government service.

(3) I am fairly satisfied with local government service, but I think it could be better in certain instances.

(4) I am mainly dissatisfied with local government service, but I think it is good in certain instances.

(5) I am completely dissatisfied with local government service.

(6) No opinion.

3. This has been "tested" by a tree analysis in which discussion frequency constituted the dependent variable. Since the discussion frequency variable consists of a trichotomy where the middle class is completely dominant, it was impossible to construct a reasonable dichotomy variable. Therefore the dependent variable was treated as an interval scale with three values. Obviously the analysis can thus be seen only as a rough indication of the plausibility of the assumption discussed above. It was evident from the tree that six of the variables had participated in the analysis.

4. The procedure is evident from the formula below:

$$\text{Res} = \sum_{i=1}^{6} \frac{n \, M_{TOT}}{100} - a$$

where n = number of interviewed persons from the respective final group in the type of commune

TOT = proportion of respondents (%) in the final groups who have a large range of demands

a = observed total number with a large range of demands in the respective type of commune

In Table 8 the residuals have subsequently been recalculated as a percentage of the total number of interviewed persons in the respective type of commune.

REFERENCES

BIRGERSSON, B.O. (1975). "Kommunen som serviceproducent: Kommunal service och serviceattityder i 36 svenska kommuner" ("The commune as a producer of services: Local government services and service attitudes in 36 Swedish communes"). Doctoral dissertation, University of Stockholm.

BRANTGARDE, L. (1971). Tradanalys—en multivariat analysteknik till-lampad for kvantitativ beroende variabel (Tree analysis—A multivariate mode of analysis). Gothenburg.

DOWNS, A. (1967). Inside bureaucracy. Boston: Little, Brown.

DYE, T. (1972). Understanding public policy. Englewood Cliffs, N.J.: Prentice-Hall.

SHARKANSKY, I. (1967). "Government expenditures and public service." American Political Science Review, (December):1066-1075.

SONQUIST, J.A. (1969). "Finding variables that work." Public Opinion Quarterly, 33(spring).

SONQUIST, J.A., and MORGAN, J.N. (1965). The detection of interaction effects. Ann Arbor, Mich.: Institute for Social Research.

TARSCHYS, D. (1975). "The growth of public expenditures: Nine modes of explanation." Scandinavian Political Studies, 10.

WESTERSTAHL, J. (1971). "Ett forskningsprogram" (A research program). Den kommunala självstyrelsen 1, Stockholm.

10

Electors and the Elected in Sweden

LARS STROMBERG

INTRODUCTION

□ A FUNDAMENTAL RATIONALE for local self-government is that solutions to locally delimited problems of value allocation should be subject to citizen influence and that their will should be realized in local political decisions and government activities.[1] In direct democracy, this goal is achieved because of the identity that prevails between citizens and decision makers. In representative democracy, the great mass of electors is distinct from decision makers and direct participation is replaced by indirect.[2]

With few exceptions, the only means of influencing the local decision-making process left to the voter in a representative democracy is participation in local general elections. In Sweden, these are the elections in which members of the communal (municipal) council are selected. Thus, the voter, through regular elections preceded by a free process of opinion formation, is given a chance to influence the council's composition. The parties compete for votes, and voters can take sides on the commune's future policies.

AUTHOR'S NOTE: *This essay was originally prepared as a paper for the ECPR Workshop on Local Government, Marstrand Sweden August 19-24, 1974.*

In this way, the voters' crucial interests are supposedly secured in decisions reached within the commune.[3]

In Sweden, this view of democracy has been strongly influenced by the doctrine of proportional list voting and the party system that has emerged within that system. Initially, the idea underlying the proportional list voting system's "reflection theory" was that the electoral system should allow various *opinions* to be represented in proportion to their strength. It was not intended that the electoral system should reflect special *class interests* within the electorate. In the mid-19th century, a clear distinction was drawn between opinions and interests.

Around the turn of the century, the two concepts came to be used as synonyms. The party system which then came into being both was class-based and also represented differing views and opinions within Swedish society. Thus, it was thought that the proportional list voting system could replace the old representation by class and estate. The difficulty inherent in the older forms of local and national representation had been that changes in society, above all the emergence of new classes at the time of the Industrial Revolution, had raised problems of how to combine interests with class representation. The class-based party system, combined with the proportional election systems, was regarded as providing a technical solution to this problem.

As the modern party system became consolidated, debate over the electoral system came to be wholly tied up with the parties' ability to be elected and represented; it is the parties who are the protagonists in the debate and who regard themselves as representing essential opinions and interests within society. However, it is reasonable to assume that the proportional list voting systems' reflection theory is still being interpreted to mean that it is the electorate's opinions or interests which must be represented and not only the parties as such.

The notion has also been entertained that the council should reflect the demographic characteristics of the electorate, since various groups within the electorate are thought to have specific interests with regard to local government activities. It is assumed that these interests are primarily represented via the party system. But in issues in which there is no divergence between one party and another or which have not been discussed in the public debate, it is assumed that the interests of various subgroups will be faithfully reflected only if they are adequately represented on the council.

ASSUMPTIONS UNDERLYING THE REFLECTION THEORY

In this paper, we examine the degree of agreement between the opinions and socioeconomic composition of the electorate and the opinions and socioeconomic composition of communal councils in Sweden. The degree of agreement between the councils' and the electorate's opinions is referred to as *opinion representativeness.* By opinion, we mean attitudes about certain widely defined fields of communal activity. The degree of agreement between the council's and the electorate's demographic composition we shall call *socioeconomic representativeness.*

Theoretically, high opinion representativeness is one of a number of conditions which must be met if the will of citizens is to be realized in political decisions and local government activities. Even if opinion representativeness is low, however, the council can still reach decisions which take prevalent opinions into account, even though council members may not share those views. Conversely, even where opinion representativeness is high, limitations on a commune's juridical competence and economic resources and, not least, lack of efficiency in its administrative system may all contribute to decisions being reached that do not meet citizen demands.

If local government elections are to function in the manner indicated, a number of other conditions must also be met. The parties must formulate and present alternatives which, in outline at least, indicate solutions to local value allocation problems and clearly distinguish between one solution and another, so as to guide the citizen in his choice of party. The voter must be well informed about what the various parties stand for in local affairs and able to identify them with a certain policy if he is to choose rationally among them.

The Swedish Local Government Research Group's studies in connection with the 1966 local elections have shown that these conditions are inadequately met. It is rare for local party organizations to participate in elections with local platforms that distinguish among them; further, the level of information among voters is low (Birgersson et al., 1971).

As a consequence, the voter does not choose among policies followed by parties within his commune but chooses his party on the basis of what he thinks the party stands for at the national level (Mansson, 1970). Thus, if local elections are to result in high opinion representativeness, the dividing lines in national politics must be reflected in attitudes toward local value allocation problems.

Opinion representativeness may also be influenced by the councils' demographic composition, as compared to that of the electorate. Among other things, individuals' demands on the local community and their attitudes toward local questions of value allocation can derive from their personal life-situations. Such a connection between political attitudes and social characteristics has long been familiar to social scientists. Different groups within the electorate are assumed to harbor different attitudes toward local value allocation questions. If the council, via the party system, does not reflect the demographic composition of the electorate, this can lead to low opinion representativeness—particularly if local government elections, via the electors' choice of parties, do not directly mirror differences of opinion among the electorate on local issues.

STUDY DESIGN

In this article, we examine a number of questions relevant to the "reflection theory" as it is discussed above. These are:

- To what extent, in local value allocation questions, are the electorate's attitudes reflected in the attitudes of council members? And how strongly does the party system affect the degree of opinion representativeness?

- How closely is the electorate's demographic composition mirrored in the councils? And how strongly does the party system affect the degree of socioeconomic representativeness?

- Is opinion representativeness influenced by variations in socioeconomic representativeness?

The data used to address these questions were obtained from two field studies undertaken by the Local Government Research Group in a sample of 36 communes, plus the city of Gothenburg. Data at the electorate level were gathered by means of individual interviews with approximately 50 voters in each commune—a total of 2,083 or 86.8% of 2,400 sampled individuals, at the time of the 1966 local elections. At the council level, data were gathered by means of a mail questionnaire in 1968, which was sent to all 1,328 councillors elected in the 1966 communal elections. Of these 1,224 or 92.1% answered the questionnaire.

The 36 sample communes were drawn using a modified "graeco-latin-square-design." As basic selection variables, size and density of

TABLE 1
THE SAMPLE OF COMMUNES

Population Density[a]	Size of Population		
	I 0-8,000 Inhabitants	II 8,000-30,000 Inhabitants	III 30,000+ Inhabitants
A. 90-100%	6 (36)	6 (42)	6 (17)
B. 30- 90%	6 (332)	6 (50)	— (1)
C. 0- 30%	6 (305)	— (2)	— (0)

a. Percentage of inhabitants living in built-up areas.

population (percentage of inhabitants living in built-up areas) were chosen. The political situation was also considered (communes where the Social Democrats did/did not have a majority in the 1964 elections) as well as geographic region. In Table 1, the commune sample construction appears in relation to the two basic variables. The numbers in parentheses indicate how many communes there actually are within the combined size-density of population classes. Six communes were taken from each class whenever possible.[4] The statistics reported are, as a rule, means of means, means of proportions, or means of correlation coefficients over the six communes in each sample cell.

OPINION REPRESENTATIVENESS

This section is devoted to an examination of variations in opinion representativeness among types of communes and political parties. Starting with the notion that politics is a question of value allocation in society, we decided to investigate two fundamental political attitudes: How extensive should local communal commitments be and how much should they cost? And how shall society, within this framework, choose its priorities as between commitments to various fields of activity? These two questions, of course, are closely connected; even so, we preferred to regard the level of society's total commitments and priorities among those commitments as two separate questions.

Attitudes toward the level of a commune's commitments or the total extent of local government activity were measured operationally in connection with attitudes about the proportional income tax. Attitudes toward a commune's commitments in various fields were operationalized in an interview question about 12 broadly defined public service areas.

With respect to the local income tax, our goal was to measure attitudes toward changes in existing tax and service levels within each commune. Where it was a question of changes in service levels in various fields, the central issue was the priority to be given to different sectors.

ATTITUDES TOWARD LOCAL EXPENDITURES (MUNIFICENCE)

Developments in Swedish society since democracy made its breakthrough in 1919 have been typified by a massive expansion of the public sector. This expansion has been a crucially divisive issue in Swedish politics—a dimension along which the political parties are distributed from right to left, the non-socialists being less positive in their attitudes toward expansion than the socialists.[5] Since the local income tax is proportional, increases in communal commitments —insofar as they are not financed with national funds—mean heavier taxes for all members of a commune.

We chose the direct link between communal commitments and local income taxes as a starting point for our measurement of attitudes toward the desirability of changes in the extent of communal commitments. In the 1966 electorate study, one of the interview questions asked about the need for communal commitments—tying this directly to the level of local income taxes. It was formulated as follows:

> The contributions made by the communes to various areas within the community do, as a rule, affect local income taxes. Which of the following standpoints do you think applies best to [name of commune]?
>
> ● There is so much to be done in this commune that the commune ought to increase its commitments, even though this would mean raising local income taxes.
>
> ● Local income taxes are on the whole satisfactory in relation to the needs of the commune.
>
> ● It ought to be possible to diminish the commitments so as to reduce local income taxes.

Exactly the same questions were put to councillors in 1968.

In order to describe variations in attitudes toward local expenditures, a simple index was constructed based on the distributions of the three alternative answers—raised, unchanged, and reduced local income taxes. These alternatives were given the values of +1, 0, and

−1 respectively and then multiplied by the percentage distributions for each alternative. Thus, the index varies from +100, which means that the whole group is in favor of raising local income taxes, to −100, implying that everyone in the group is in favor of a tax reduction.[6]

In Table 2, the attitudes of citizens and council members are compared in percentages and in index scores calculated in the manner described above. Two very clear tendencies can be discerned. First, councillors in all types of communes are a great deal more positive toward local expenditures (more munificent) than the electorate. Second, munificence rises among councillors with size of commune, while the opposite is true of voters.

Differences among commune types in point of munificence as between councils and the electorate indicate that opinion representativeness is lower in the larger communes. This is evident from the measure of opinion representativeness which is also presented in Table 2. The index of opinion representativeness is the difference between councillors and citizens in the munificence index (irrespective of sign) divided by 200, which is the maximal difference that the munificence index can show. This measure of concurrence varies between 0 (which means that the munificence index for councillors

TABLE 2

ATTITUDES TOWARD LOCAL EXPENDITURES AMONG COUNCILLORS AND VOTERS IN VARIOUS TYPES OF COMMUNES. PERCENTAGE AND INDEX.

Population Density[a]		Size of Population			
		I 0-8,000	II 8,000-30,000	III 30,000+	Gothenburg
	Munificence index				
	Councillors	+13	+38	+34	+75
A. 90-100%	Voters	− 2	− 9	−18	−22
	Index of opinion representativeness	0.08	0.23	0.26	0.48
	Munificence index				
	Councillors	+27	+28		
B. 30- 90%	Voters	+ 5	− 7		
	Index of opinion representativeness	0.11	0.18		
	Munificence index				
	Councillors	+15			
C. 0- 30%	Voters	+ 8			
	Index of opinion representativeness	0.04			

a. Percentage of inhabitants living in built-up areas.

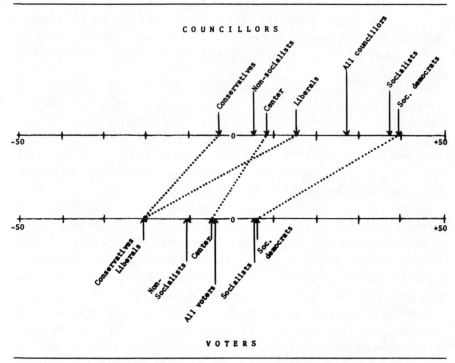

Figure 1: ATTITUDES TOWARD EXPENDITURES AMONG COUNCILLORS AND VOTERS OF DIFFERENT PARTIES (Average Munificence Scores)

and voters agrees perfectly) and 1 (which means that it is maximally different).

In the previous section, we hypothesized that munificence reflected the right-left dimension in local politics. We tested this hypothesis by investigating munificence among various parties within the electorate and within the councils. A crucial question in this context is: Are the various parties in the councils representative of their respective electorates?

Munificence, measured in terms of the above index, is shown in averages for all 36 communes and for each of the four major parties—the Conservatives, the Centre party, the Liberal party, and the Social-Democratic party—in Figure 1.

At the electorate level, munificence is lowest in the Conservative and Liberal parties, followed by the Centre party. All of these parties show negative index scores—which is to say, their average voter would like to reduce communal commitments and local income taxes. The same holds true, albeit to a lesser extent, of the few Communists in the electorate sample. The Social-Democratic electors show a faintly positive index, which indicates that, as a group, they are somewhat more munificent than people voting for the other parties. The differences are, however, not large.

At the council level, the same rank order is found among the parties as in the electorate, apart from the fact that the Centre and Liberal parties have exchanged positions. At the council level, however, only Conservative party councillors show any desire to diminish communal commitments and local income taxes. The Centre, the Liberals, and, particularly, the Social Democrats desire an increase. Councillors of all parties are a good deal more munificent than those who elect them. The difference between councillors and voters is particularly notable within the Social-Democratic party.

These findings are summarily presented in Table 3, where the munificence index for nonsocialist and socialist councillors and electors is presented for the various commune types. Everywhere socialist electors are more munificent than nonsocialists—a difference which, however, is smaller in the larger communes. Socialist councillors are, similarly, more munificent than nonsocialists but the differences among councillors of different parties increase with the size of commune. As a result, socialist councillors, as the size of commune increases, tend to differ more from their electors than do nonsocialist councillors from theirs. The negative association between opinion representativeness and size of commune thus appears to be due largely to the fact that socialists in the councils of larger urban communes deviate sharply in their attitudes toward expenditures from those who have voted for them. This is evident from the Index of Opinion Representativeness also shown in Table 3.

ATTITUDES TOWARD COMMUNAL SERVICE PRIORITIES

The measurement of priorities among various kinds of local public services is fraught with theoretical and practical difficulties. The first of these is how to construct an interview question which covers a wide range of local public activities. After a test study, in connection with the Research Group's information study at the time of the 1966 local elections when local politicians were interviewed about current local issues, a list of such issues was drawn up. This list was used as a basis for distributing local government activities into the following 12 service fields found in all types of communes:[7]

TABLE 3

INDEX OF ATTITUDES TOWARD LOCAL EXPENDITURES (MUNIFICENCE) AMONG NONSOCIALISTS AND
SOCIALISTS AND INDEX OF OPINION REPRESENTATIVENESS IN VARIOUS TYPES OF COMMUNES

| | | Size of Population | | | | | | |
| | | I 0-8,000 | | II 8,000-30,000 | | III 30,000+ | | Gothenburg | |
Population Density[a]	Respondents	Non-socialists	Socialists	Non-socialists	Socialists	Non-socialists	Socialists	Non-socialists	Socialists
A. 90-100%	Councillors	+10	+15	+ 4	+65	+ 2	+66	+51	+100
	Voters	− 8	+ 7	−24	+ 1	−21	−12	−16	− 11
	Index of opinion representativeness	0.09	0.08	0.14	0.32	0.12	0.39	0.34	0.56
B. 30- 90%	Councillors	+ 7	+43	− 7	+49				
	Voters	+ 9	+ 8	−16	+ 6				
	Index of opinion representativeness	0.01	0.18	0.05	0.22				
C. 0- 30%	Councillors	+ 5	+31						
	Voters	−10	+24						
	Index of opinion representativeness	0.08	0.03						

a. Percentage of inhabitants living in built-up areas.

Town planning

Housing

Employment opportunities

Streets, roads, and traffic

Waterworks and sewage

Schools

Day nurseries

· Home help for families with children and the elderly

Old-age homes

Social assistance

Leisure activities

Libraries and other cultural activities

In our interviews with voters, interviewees were asked whether in each of the 12 fields they thought "more should be done," "things are good as they are," the commune's commitments "could be reduced," or whether they "had no opinion." Answers were given for each service field in connection with a questionnaire card.[8]

With the aid of answers from the electorate, we constructed a rank-ordering of service fields. To do this, a mean, based on the first three of the above possible answers ("more should be done," "things are good as they are," "could be reduced") was calculated for the whole electorate (or various subgroups) for each of the 12 items on the assumption that these three alternatives could be treated as an interval scale with the values of +1, 0, and −1. Those persons who gave no opinion were scored with the intermediate alternative. These mean values were then used to rank order the 12 service fields within each commune.[9] The service question was repeated in the 1968 council questionnaire—phrased, to the extent possible, like the question in the 1966 voter interviews.

If we look, first, at what priorities voters give to the various types of services, libraries, schools, and social assistance usually come at the bottom of the list in all types of commune (no table shown). In the intermediate group, we find old-age homes, leisure activities, and town planning. High priority is usually given to housing and home help. There are, however, some differences among commune types. While housing and home help, for example, are high-ranking service fields in urban communes, employment is given high priority in small communes. Another example is institutionalized care of the aged, which is always ranked higher in urban than in sparsely populated communes.

Differences in priorities derive from essentially different types of social problems, as between town and country. At the time when the study was made, housing shortage was a severe problem in urban areas, just as employment was in the sparsely populated rural communes. In either case this is probably an expansion of maladaptation to the swift urbanization of Swedish society in recent years. The demand for more home help and for an extension of institutionalized care of the aged in urban communes can probably be derived from the same phenomenon.

In order to obtain a comprehensive measure of these differences in service priorities, Spearman's rank correlation coefficient has been calculated pairwise for the six sample cells and Gothenburg.[10]

These correlation coefficients have been grouped together in a matrix (Table 4), which shows clearly that, on the one hand, considerable differences exist between the voters' rank-orderings in the small sparsely populated rural communes and in Gothenburg. On the other hand, the rank-orderings are very close in the larger urbanized communes (cell IIIA) and in Gothenburg. As can be seen, a systematic connection exists, by and large, between the concurrence in rank-ordering and in commune structure.

The same types of variations are found at the council level. The councillors' rank-orderings, however, differ from those of the electorate, particularly in urbanized communes. In all commune types, three fields especially stand out. These are employment, town planning, and water, which are always ranked higher—in urban communes, considerably higher—in the councils than among the

TABLE 4

CORRELATION MATRIX FOR PAIRWISE COMPARISONS BETWEEN ELECTORS' RANK-ORDERING OF 12 SERVICE FIELDS IN VARIOUS TYPES OF COMMUNES. SPEARMAN'S RHO

		Type of Commune						
		IC	IB	IIB	IA	IIA	IIIA	Gothenburg
	IC	—	0.92	0.94	0.85	0.75	0.77	0.49
	IB	—	—	0.88	0.79	0.70	0.72	0.45
Type	IIB	—	—	—	0.89	0.87	0.90	0.70
of	IA	—	—	—	—	0.82	0.87	0.66
Commune	IIA	—	—	—	—	—	0.98	0.88
	IIIA	—	—	—	—	—	—	0.91
	Gothenburg	—	—	—	—	—	—	—

TABLE 5

SPEARMAN'S RANK ORDER CORRELATION COEFFICIENT BETWEEN
COUNCILLORS' AND VOTERS' RANKINGS OF SERVICES

	Size of Population			
Population Density[a]	I 0-8,000	II 8,000-30,000	III 30,000+	Gothenburg
A. 90-100%	0.56	0.45	0.64	0.56
B. 30- 90%	0.71	0.76		
C. 0- 30%	0.75			

a. Percentage of inhabitants living in built-up areas.

voters. It would seem reasonable to suppose that voters generally regard these issues as less urgent and do not see that housing shortages, available land, and town planning are all linked together.

To facilitate comparisons among council members and the electorate, Spearman's rank-order correlation coefficients were calculated for *every* commune. The means for these coefficients are shown in Table 5. As indicated, rather clear differences exist among commune types. Concurrence in rank orderings is somewhat higher in very sparsely populated rural communes, communes of a purely rural character, and communes of the mixed urban and rural type than in the purely urbanized ones. The coefficients, however, are high throughout and thus indicate a rather good concurrence between the voters' and the representatives' service priorities.

In order to determine whether or not there are party differences in the degree of concurrence between councillors and electors in their ordering of service priorities, Spearman's rank-order correlation coefficients were calculated separately for the socialist and nonsocialist parties. These coefficients show very small differences between the two blocks in most commune types. Earlier we have seen that differences between the councils as a whole and the electorate as a whole in service priorities can be derived from a different rank-ordering of service fields which have a strong planning element. This tendency is also found when opinion representativity is compared with the socialist and the nonsocialist blocks (no table shown).

SOCIOECONOMIC REPRESENTATIVENESS

Two questions are addressed in this section. The first is whether socioeconomic representativeness varies with commune type. The second is how well do councillors of different parties reflect their own segments of the electorate. Socioeconomic representativeness is here defined as a measure of the degree to which councillors reflect the demographic composition of the electorate in several different, politically relevant respects including sex, age, and occupational and social stratification.

SEX AND AGE

In Swedish public debate, the sex and age distributions in councils have been the most discussed of all socioeconomic characteristics. The number of women elected has always been low, although campaigns for better representation have increased the share of women representatives from about 10% to 17% in the last 20 years. When this study was conducted (1968) the average share of women in the councils was 15% with little variation among different commune types. Since women make up about half of the population in all communes, they are grossly underrepresented.

Before the nominations for the 1966 elections, there was also intense agitation among political youth organizations for younger candidates. This campaign's effect, however, was limited, and there are striking differences between the ages of councillors and voters —not exclusively by virtue of underrepresentation of younger voters.

In all types of communes, voters aged 40 to 64 tend to be overrepresented while voter groups under and over these limits are, as a rule, underrepresented. The very youngest age-groups (the approximately 10% to 15% who are under the age of 30) and, likewise, the very oldest (the 15% of the population who are over 70) are virtually unrepresented in the councils (no tables shown).

SOCIAL CLASS

It is generally assumed that occupation is a measure of social class.[11] In the analyses that follow, we relied upon the social class classification used by the National Central Bureau of Statistics. In this classification, Social Class II consists of large scale entrepreneurs, managers of firms, senior white-collar workers, and certain self-employed persons. Social Class II comprises all other white-collar workers and self-employed persons, including farmers. Social Class III consists wholly of blue-collar workers.

This distribution into social classes has been criticized for its lack of theoretical foundation. We find no clear norms upon which the distribution, allegedly based on economic and social criteria, is constructed. Nevertheless, this classification has proved useful in explaining political attitudes and behavior. One special problem is the difficulty of classifying married women working in the home. Housewives make up a large group in the electorate, and where they should be placed in the classification is unclear. We have opted for the traditional solution: to code them according to their spouse's social class.

In Table 6 the percentage of electors and councillors in each of these three social classes is shown together with an index of socioeconomic representativeness. This index is the ratio of a class's percentage share of the councillors to its share of the electorate, multiplied by 100. This yields a measure which expresses the percentage of councillors from any given class—compared to what that percentage would be if it were proportional to the group's strength in the electorate. The value 0 indicates that zero percent of a given group of voters is represented in the councils; the value 100 indicates that the percentage of councillors from a class is the same as the percentage of voters in that class; and values larger than 100 indicate an overrepresentation of the group in question.[1][2]

It is evident both from the percentage differences and the representativeness index that Social Class III is strongly underrepresented and that this underrepresentation increases with size of commune. Both Social Classes II and I are overrepresented—Social Class I to an exceptionally high degree.

Overrepresentation of Social Classes I and II varies with commune type. Social Class II's overrepresentation decreases with size of commune while Social Class I's overrepresentation increases with size of commune. Although Social Class II dominates the electorate in all types of communes, Social Class II is the largest group in the councils with the exception of Gothenburg, where Social Class I assumes this role.

An alternative measure of social stratification is education. Education shows the same tendencies as social class—an expression of the strong mutual relationship between these two variable. Thus, as the communes become larger and more urbanized, their councils are progressively dominated by a well-educated middle class (no table shown).

TABLE 6

**SOCIAL CLASS DISTRIBUTIONS AMONG COUNCILLORS AND VOTERS,
MEAN PERCENTAGE AND INDEX OF REPRESENTATIVITY (RI): BY TYPE OF COMMUNE**

	Size of Population															
	I 0-8,000 Social Class				II 8,000-30,000 Social Class				III 30,000+ Social Class				Gothenburg Social Class			
Population Density[a]	I	II	III	Total %	I	II	III	Total %	I	II	III	Total %	I	II	III	Total %
A. 90-100%																
Councillors	20	53	27	100	24	52	24	100	38	44	18	100	52	38	10	100
Voters	7	41	52	100	6	42	52	100	9	39	52	100	9	40	51	100
Difference	+13	+12	−25	0	+18	+10	−28	0	+29	+5	−34	0	+43	−2	−41	0
RI	297	128	51		384	125	46		413	111	36		590	94	20	
B. 30-90%																
Councillors	11	54	35	100	23	49	28	100								
Voters	2	33	65	100	2	38	60	100								
Difference	+9	+21	−30	0	+21	+11	−32	0								
RI	473	164	56		1140	130	47									
C. 0-30%																
Councillors	6	64	30	100												
Voters	2	43	55	100												
Difference	+4	+21	−25	0												
RI	305	150	35													

a. Percentage of inhabitants living in built-up areas.

INCOME

Each representative in the elected council combines a number of demographic characteristics in multidimensional space. We felt a need to reduce this multidimensional space, if possible, to a single dimension. Technically this ought to be possible by means of some kind of scaling method. A simpler method is to make direct use of such a comprehensive measure—namely, income.

A simple comprehensive measure of income at the voter and the councillor levels is median income (Table 7). This shows the same tendencies as social class—naturally enough in view of the strong correlation between these two variables. Differences in income between councillors and voters are considerable—councillors having incomes two to three times as great as those of the electorate. These differences increase with size of commune and degree of urbanization.

Income distributions among councillors and the electorate were also compared by means of cumulative income distributions. From these income distributions, we calculated Gini indices for each of the commune types. In a Gini index, the area between the so-called Lorentz curve and the line of equal representation is divided by the whole area beneath the line of equal representation. If the Lorentz curve coincides with the line of equal representation, the Gini index is 0. The more the Lorentz curve deviates from the line of equal

TABLE 7
MEDIAN INCOME AMONG COUNCILLORS AND VOTERS:
BY TYPE OF COMMUNE (Swedish Kronas Per Annum)

Population Density[a]	Size of Population			
	I 0-8,000	II 8,000-30,000	III 30,000+	Gothenburg
A. 90-100%				
Councillors	26.900	27.200	32.000	44.200
Voters	10.200	11.700	11.700	12.200
Difference	+16.700	+15.500	+20.300	+32.000
B. 30- 90%				
Councillors	20.200	24.000		
Voters	6.700	8.900		
Difference	+13.500	+15.100		
C. 0- 30%				
Councillors	19.400			
Voters	7.900			
Difference	+11.500			

a. Percentage of inhabitants living in built-up areas.

representation, i.e., the greater the differences between the distributions of the two compared populations, the closer the index value will be to 1. Where the two populations are completely unalike, the Lorentz curve will coincide with the x and y axes in which case the Gini index will yield a value of 1.[13]

The means for the Gini index in various commune types are presented in Table 8. There is, generally speaking, both an urbanization and a size effect. The larger the commune and the higher its degree of urbanization, the more the councillors' income distributions deviate from those of the electorate.

VARIATIONS AMONG PARTIES

The degree of socioeconomic representativeness in each commune depends on how well the various parties reflect those segments of the electorate which vote for them. A number of studies at the electorate level, notably in connection with parliamentary elections, indicate that differences do exist in the occupational and social composition of the parties. The Communists and the Social Democrats are dominated by working-class voters; the Centre party finds support chiefly among farmers; and the Conservatives and Liberals recruit their voters chiefly from among white-collar workers and entrepreneurs (Särlvik, 1974). There are, however, few differences among parties with respect to such variables as sex and age.

In our own study, we also found little difference among the various parties within the electorate with respect to sex and age. At the councillor level, there are minor differences in age composition, but these are, generally speaking, negligible. Throughout, women are weakly represented on local councils, and this is true for all parties.

There are, however, differences in the electorate and in the councils in the social class composition of the parties. Among voters,

TABLE 8

**GINI INDEX FOR INCOME DISTRIBUTION COMPARING COUNCILLORS'
AND VOTERS' INCOMES IN VARIOUS TYPES OF COMMUNES**

	Size of Population			
	I	II	III	
Population Density[a]	0-8,000	8,000-30,000	30,000+	Gothenburg
A. 90-100%	0.65	0.62	0.71	0.75
B. 30- 90%	0.58	0.65		
C. 0- 30%	0.52			

a. Percentage of inhabitants living in built-up areas.

the nonsocialist blocks are dominated by Social Class II (between 50% and 70% in the various commune types) and the socialist blocks, by Social Class III (between 67% and 80%). The social class composition of the parties also varies with commune type. The proportion of voters in Social Class II among the socialists and Social Classes I and II among nonsocialists increases with commune size and density. The same relationships exist between social class and commune type among the councils but are even more marked (Table 9).

These differences in social class distributions mean that the representativeness index for each social class will also vary by party. In the councils, the nonsocialists are strongly underrepresentative of their Social Class III voters and overrepresentative of Social Classes I and II—particularly in the larger, more urbanized communes. The socialists show the same tendencies but are less underrepresentative of Social Class III than nonsocialists and more overrepresentative of Social Classes II and I.

OPINION AND SOCIOECONOMIC REPRESENTATIVENESS

In this section, we examine the relationship between socioeconomic representativeness and opinion representativeness. We look first at the relationship between socioeconomic representativeness and "munificence" and next, at the relationship between socioeconomic representativeness and attitudes toward service priorities.

SOCIOECONOMIC REPRESENTATIVENESS AND ATTITUDES TOWARD LOCAL EXPENDITURES (MUNIFICENCE)

Earlier, we noted that councillors tend to be more munificent than voters and that these differences increase with commune size. The Gini index of income differences between councillors and the electorate also shows a strong connection with commune size. Is there, then, a statistical relationship between socioeconomic representativeness and opinion representativeness? And, if so, can it be an expression of some causal connection?

To answer the first question, the concurrence in income distributions between councillors and the electorate, measured by the Gini index, were correlated with opinion representativeness for munificence in the sample communes. This analysis revealed a positive

TABLE 9

SOCIAL CLASS DISTRIBUTION AMONG NONSOCIALISTS AND SOCIALISTS—MEAN PERCENTAGE AND INDEX OF REPRESENTATIVITY (RI): BY TYPE OF COMMUNE

Size of Population

Population Density[a]	I 0-8,000 Social class				II 8,000-30,000 Social class				III 30,000+ Social class				Gothenburg Social class			
	I	II	III	Total %	I	II	III	Total %	I	II	III	Total %	I	II	III	Total %
Non-socialist																
A 90-100%																
Councillors	30	63	7	100	43	53	4	100	55	41	4	100	64	37	0	100
Voters	10	54	36	100	12	69	19	100	16	55	29	100	18	57	26	100
Difference	+20	+9	-29	0	+31	-16	-15	0	+39	-14	-25	0	+46	-20	-26	0
RI	300	117	19		358	77	21		344	75	14		356	65	0	
B 30-90%																
Councillors	19	69	12	100	33	63	4	100								
Voters	6	61	33	100	6	45	49	100								
Difference	+13	+8	-21	0	+27	+18	-45	0								
RI	317	113	36		550	140	8									
C 0-30%																
Councillors	9	85	6	100												
Voters	1	61	38	100												
Difference	+8	+24	-32	0												
RI	900	139	16													
Socialists																
A 90-100%																
Councillors	6	46	48	100	7	51	42	100	21	47	32	100	41	38	21	100
Voters	3	25	72	100	2	28	70	100	4	29	67	100	2	21	77	100
Difference	+3	+21	-24	0	+5	+23	-28	0	+17	+18	-35	0	+39	+17	-56	0
RI	200	184	67		350	182	60		525	162	48		2050	181	27	
B 30-90%																
Councillors	4	41	55	100	9	42	49	100								
Voters	1	12	87	100	1	30	70	100								
Difference	+3	+29	-32	0	+8	+12	-21	0								
RI	400	342	63		900	140	70									
C 0-30%																
Councillors	1	29	70	100												
Voters	0	19	80	100												
Difference	+1	+10	-10	0												
RI	100	153	88													

[a]Percent of inhabitants living in built-up areas.

relationship between the two indices. The lower the socioeconomic representativeness, the greater the differences in munificence between councillors and the electorate. The proportion of explained variance amounted to 24% (no table shown).

The greater the difference between the incomes of councillors and voters, then, the more munificent are the councillors. In view of the relationship, often demonstrated in social science literature, between right-wing attitudes and high social status and between left-wing attitudes and low social status, this is a peculiar result (Lipset, 1959; Key, 1961). One explanation of the result might be that, compared with progressive national taxes, the proportional local income taxes favor higher income brackets and that it would thus be in the interests of persons in higher income brackets to keep local taxes at a high level. This interpretation seems unlikely, however, since socialist councillors are considerably more munificent than their nonsocialist colleagues—particularly in the larger communes—even though their own incomes are lower.

We have reason to suspect, therefore, that the relationship between socioeconomic and opinion representativeness may well be an artifact of party membership. Correlations between the Gini index for income and opinion representativeness for munificence were also calculated separately for socialist and nonsocialist parties. This comparison revealed a considerably lower relationship between socioeconomic representativeness and opinion representativeness for both blocks than when the electorate as a whole was compared with all councillors. The proportion of explained variance approached 0 for the nonsocialists and amounted to 6% for the socialists. Thus, it is only for the latter that a faint connection exists between socioeconomic and opinion representativeness.

Thus, the connection between socioeconomic representativeness (in terms of income) and opinion representativeness is far from clear. In Figure 2, we show the relationships between income and munificence both for voters and councillors. There the average munificence index for five income classes is shown for all 36 communes. As indicated, the relationships between income and munificence are not linear either at the councillor or the electorate level. Rather they take a curvilinear form, implying that munificence both among councillors and electors is low in the lower income brackets, rises to a peak, and then falls again in the very highest income brackets. The relationships at the councillor and voter levels have approximately the same form but differences between the two

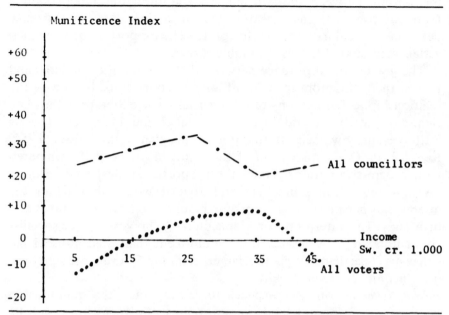

Figure 2: **INDEX OF ATTITUDES TOWARD LOCAL EXPENDITURES:**
BY INCOME GROUPS (Mean Index Score for 36 Communes)

levels are rather large. Everywhere, councillors are more munificent in all income classes than are voters.

Not shown in Figure 2 are the considerable differences that exist among commune types. The simplest way of describing these is to say that the smaller the commune, the closer the two curves lie to one another, and the larger the commune, the farther apart they lie. Differences among commune types are due primarily to the councillors' attitudes. Even among the highest income councillors in the larger communes, munificence is rather high. At the electoral level, the relationships between income and munificence assume approximately the same form in all commune types.

The income of councillors, thus, appears to exert little influence on munificence. This conclusion is not greatly modified when councillors and voters are divided into socialists and nonsocialists. In Figure 3, we see that the form of the relationship between munificence and income for nonsocialist and socialist voters is relatively similar even if the socialists are, on the average, somewhat more munificent. On the other hand, socialist and nonsocialist councillors are unalike both in terms of the level of munificence and the form of the relationship between munificence and income. For

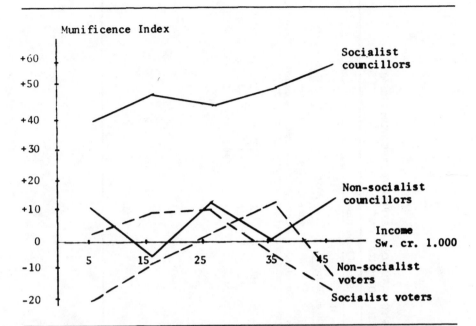

Figure 3: INDEX OF ATTITUDES TOWARD LOCAL EXPENDITURES (MUNIFICENCE): BY INCOME GROUP AND PARTY ORIENTATION

the nonsocialists, there appears to be no relationship between munificence and income; for the socialists there is a weak positive and virtually rectilinear one. This relationship becomes even more rectilinear with rising commune size.

From the foregoing, it is clear that a change in the income composition of councils would have relatively little impact on "munificence" among nonsocialist councillors. A change in the direction of better socioeconomic representativeness among socialists, on the other hand, would probably lead to a reduction in munificence and, therefore, an increase in opinion representativeness. This is particularly true for larger communes.

One method of measuring the bearing of inadequate socioeconomic representativeness upon the attitude distribution of councillors with respect to munificence is to standardize the attitude distributions in councils with the income distribution in the electorate. In Table 10, we show, for various commune types, the observed munificence values for councillors and the electorate and the predicted index values for councillors if their income distribution had been the same as that of the electorate. It is evident from this table that both the socialist and the nonsocialist councillors'

TABLE 10

OBSERVED AND PREDICTED INDEX SCORES FOR ATTITUDES TOWARD LOCAL EXPENDITURES (MUNIFICENCE): BY TYPE OF COMMUNE AND PARTY ORIENTATION[a]

| | Size of Population | | | | | | | |
| | I 0-8,000 | | II 8,000-30,000 | | III 30,000+ | | Gothenburg | |
Population Density[b]	Non-socialists	Socialists	Non-socialists	Socialists	Non-socialists	Socialists	Non-socialists	Socialists
A. 90-100%								
Observed index score—councillors	+10	+15	+ 4	+65	+ 2	+66	+51	+100
Observed index score—voters	− 7	+ 7	−24	+ 1	−21	−12	−16	− 11
Predicted index score	+ 4	+15	−18	+59	+29	+46	+51	+100
B. 30- 90%								
Observed index score—councillors	+ 7	+43	− 7	+49				
Observed index score—voters	+ 8	+10	−16	+ 6				
Predicted index score	+10	+46	− 4	+53				
C. 0- 30%								
Observed index score—councillors	+ 5	+31						
Observed index score—voters	−10	+24						
Predicted index score	+15	+35						

a. Predicted index scores based on income distributions in the electorate.
b. Percentage of inhabitants living in built-up areas.

predicted and actual indices are very similar.[14] It seems, then, that the characteristic of being a councillor and belonging to a certain party exerts more influence on munificence among councillors than does social status, here operationalized as "own income."

The findings with respect to other socioeconomic characteristics are similar to those for income. Sex, age, and social class all show very weak relationships with munificence—at both the councillor and the electorate levels. Like income, these relationships are nonlinear. In Table 11, we show observed and predicted values for munificence (calculated in the same ways as those for income in Table 10) to determine how councillors' munificence would be altered if the composition of the councils were changed with respect to these variables.

As indicated, the demographic composition of councils appears to have limited bearing on councillor attitudes toward communal expenditures. The only major deviation occurs in cells II and III, where the predicted values for councillors' munificence are higher than the observed values. These differences are due largely to the fact that the nonsocialist parties in cells II and III strongly underrepresent social group III. In the councils, this social group is dominated by munificent Social Democrats and Communists. Otherwise, the differences between observed and predicted index values are small.

It is possible that munificence is higher in the councils than in the electorate because of the councils' role in the political process as compared to that of the electorate. The individual voter probably thinks about his particular needs and demands on societal commitments and may not see the relationship between his demands and communal expenditures and taxes. Councillors, on the other hand, represent a variety of interests and need to consider a wide spectrum of demands.

SOCIOECONOMIC REPRESENTATIVENESS AND ATTITUDES TOWARD COMMUNAL SERVICE PRIORITIES

Earlier, we noted that a relatively high degree of concurrence exists among voters and councillors in their attitudes toward service priorities. We also noted, however, that opinion representativeness with respect to service priorities varies among commune types. Opinion representativeness is lower in the densely populated urban communes than in either the extremely sparsely populated rural ones or those of mixed character. The analysis of socioeconomic

TABLE 11

OBSERVED AND PREDICTED INDEX SCORES FOR ATTITUDES TOWARD
LOCAL EXPENDITURES: BY TYPE OF COMMUNE AND
SOCIOECONOMIC CHARACTERISTICS[a]

	Size of Population			
	I	II	III	
Population Density[b]	0-8,000	8,000-30,000	30,000+	Gothenburg
A. 90-100%				
Observed index score				
—councillors	+13	+37	+35	+75
Predicted index score:				
sex	+ 7	+26	+33	+75
Predicted index score:				
social class	8	+48	+43	+57
Predicted index score:				
income	+ 8	+33	+39	+77
Observed index score				
—voters	− 2	− 9	−18	−22
B. 30- 90%				
Observed index score				
—councillors	+27	+28		
Predicted index score:				
sex	+37	+30		
Predicted index score:				
social class	+33	+36		
Predicted index score:				
income	+24	+30		
Observed index score				
—voters	+ 4	− 7		
C. 0- 30%				
Observed index score				
—councillors	+15			
Predicted index score:				
sex	+25			
Predicted index score:				
social class	+19			
Predicted index score:				
income	+23			
Observed index score				
—voters	+ 7			

a. Predicted index scores based on distributions of these socioeconomic characteristics in the population.
b. Percentage of inhabitants living in built-up areas.

representativeness revealed a similar relationship between commune type and socioeconomic representativeness. To what extent, we may ask, does the lack of socioeconomic representativeness explain differences between councillors and voters in their ranking of service priorities?

In order to answer this question, we first computed correlation coefficients between the Gini index for councillors' and voters' income distributions and the Spearman's rank-order correlation coefficients for service priorities. This analysis indicated that there is relatively little relationship between socioeconomic representativeness and opinion representativeness with respect to service priorities—the amount of explained variation amounting to little more than 2%. A division into socialists and nonsocialists showed the same result (no tables shown). What relationship does exist appears to be slightly negative: the less a council's income distribution agrees with the electorate's, the higher is its opinion representativeness.

We next looked at the Spearman's rank-order correlation coefficients between service priorities for councils as a whole and service priorities among high and low income voters. In calculating these coefficients, we first divided the electorate into high and low income groups and then calculated separate correlation coefficients between the service priorities of voters for each group and those of the whole council. High and low income groups were defined on the basis of the extent to which income groups were either under- or overrepresented in different commune types. As we have seen, low income brackets are underrepresented and high income brackets, overrepresented.

If the councillors also systematically overrepresent "high income attitudes," we sould expect a higher correlation coefficient when councillors' service priorities are compared to high income voters' priorities than when they are compared to the priorities of low income voters. But such is not the case (Table 12). Rather, the councillors' service priorities, particularly in urbanized communes (with the exception of Gothenburg) show higher rank-order correlation coefficients for low income brackets than high income brackets. We also found the same trend when the two party blocks were treated separately (no table shown). In both the nonsocialist and socialist blocks, councillors' service priorities were more similar to those of low income voters than to those of high income voters. Although the differences are not large, they persist throughout all commune types and party blocks.

The relationships between opinion representativeness with respect to service priorities and other socioeconomic characteristics were also examined. With respect to sex, we found that even though women are strongly underrepresented in councils, their priorities show equally good or somewhat better agreement with councillors'

TABLE 12

SPEARMAN'S RANK ORDER CORRELATIONS BETWEEN COUNCILLORS AND VOTERS OF VARIOUS INCOME GROUP'S RANKING OF SERVICE FIELDS: BY TYPE OF COMMUNE

| | Size of Population | | | |
Population Density[a]	I 0-8,000	II 8,000-30,000	III 30,000+	Gothenburg
A. 90-100%				
Low-income councillors and voters	0.59	0.48	0.69	0.56
High-income councillors and voters	0.46	0.39	0.49	0.58
B. 30- 90%				
Low-income councillors and voters	0.72	0.70		
High-income councillors and voters	0.69	0.72		
C. 0- 30%				
Low-income councillors and voters	0.68			
High-income councillors and voters	0.67			

a. Percentage of inhabitants living in built-up areas.

rank-orderings than the priorities of male voters (no table shown). In terms of age, the youngest voters (those between 23 and 34 years of age) and the oldest voters (those over 65) tend to be underrepresented while those in the 35 to 65 age group tend to be overrepresented. The rank-order correlation coefficients between councillors' priorities and those of the three age groups are shown in Table 13.

The differences between the rank-order correlation coefficients for the various age groups are relatively small. The correlation coefficients for the youngest voters are about the same as those for the intermediate group. On the other hand, the oldest voters tend to have lower rank-order coefficients when their service priorities are compared to those of the councils than do the younger and intermediate age groups. This was true for 26 of the 36 communes studied—the exceptions being the larger urban districts. Weak representation of older voters in the councils may thus lead to their special views on communal services not being heard.

With respect to social class, we concentrated on the effects of strong underrepresentation of Social Class III in the councils. Using an analytical framework similar to that reported for income, age, and sex, we found little relationship between social class representation and communal council attitudes with respect to service priorities.

TABLE 13
SPEARMAN'S RANK ORDER CORRELATIONS BETWEEN COUNCILLORS'
AND VOTERS' RANKING OF SERVICE FIELDS:
BY TYPE OF COMMUNE AND AGE GROUP

	Size of Population			
	I	II	III	
Population Density[a]	0-8,000	8,000-30,000	30,000+	Gothenburg
A. 90-100%				
23-34 years of age	0.62	0.46	0.67	0.53
35-65 years of age	0.50	0.45	0.60	0.52
66 years and older	0.42	0.40	0.63	0.57
B. 30- 90%				
23-34 years of age	0.64	0.72		
35-65 years of age	0.70	0.73		
66 years and older	0.63	0.57		
C. 0- 30%				
23-34 years of age	0.57			
35-65 years of age	0.73			
66 years and older	0.45			

a. Percentage of inhabitants living in built-up areas.

In summary, then, there appears to be little relationship between socioeconomic and opinion representativeness. Attitudes toward communal service priorities are, in other words, affected more by the characteristic of being a councillor than the council's socioeconomic composition.

SUMMARY

This article has focused on the question of how representative elected councils are with respect to matters concerning the allocation of local values (opinion representativeness) and with respect to the demographic composition of the electorate (socioeconomic representativeness).

The study of opinion representativeness focused on two fundamental attitudes toward local public activities: attitudes toward local taxes and expenditures and priorities among 12 different public service fields. The analysis of attitudes toward local public expenditures—which can be taken as an indication of the right-left dimension in local politics—showed that councillors were much more disposed to increase expenditures (be munificent) than were voters. This lack of opinion representativeness was greater in large communes than in small and in urban than in sparsely populated rural areas.

With respect to public service priorities, relatively good agreement prevailed on the whole between voters and councillors. When the rank-orderings of the various service fields are compared for the two groups, the correlations, with a few exceptions, are found to be significant. It is clear, however, that policy agreement is somewhat higher in rural than in urban areas.

The main result of the study of socioeconomic characteristics was that councillors differ significantly from voters. Middle-class people, those with high income and education, and men and people in their forties and fifties are all overrepresented on the councils. With respect to social class, income, and education, this overrepresentation is greatest in the larger and more urbanized communes.

Although variations among members of different party groups in socioeconomic characteristics reflect variations within the electorate as a whole, it is evident that councillors of all parties—particularly in the large communes—enjoy a higher socioeconomic status than their voters. This lack of socioeconomic representativeness stems from the system of nominating candidates within the party organizations. In these nominations, high priority is given to people who take active part in local politics. Such activities tend to be associated with higher socioeconomic status.

As a result, women, young people, old-age pensioners, and blue-collar workers are all largely eliminated from the nomination process. This bias in the selection process increases with size of commune, and, therefore, the group of prospective candidates differs more from the electorate in these communes than in the smaller communes (Strömberg, 1974a).

The definitive test—according to modern elite theories—of whether or not an elite can be said to exist is whether there are differences in preferences among the voters, whether the political leaders represent these diverse preferences, whether these leaders form a tightly knit group, and whether their particular interests are served by political decisions. An analysis along these lines shows that members of local councils in Sweden can hardly be regarded as an "elite."

First, if we look at electoral attitudes toward local expenditures and taxes, correlation with social class (as measured by income) is weak. The willingness to increase public expenditures is low in the lower income brackets, rises to a peak somewhere around average income, and is again low for the high income groups. Although weaker, the same type of relationship exists among councillors.

More importantly, however, councillors are much more willing to spend public funds, regardless of their own personal income and social status, than is the electorate. The same tendency is found in all parties. The level of acceptable expenditures does, however, differ among parties—socialist councillors and voters being more willing to support increased public expenditures than nonsocialists.

These results indicate that the relationship between opinion representativeness and commune type cannot be explained by variations in socioeconomic representativeness. Further analysis indicates that variation in voter participation in local politics is a more likely explanation of the differences in opinion and socioeconomic representativeness among different commune types. The participation level (i.e., the proportion of voters who are members of political parties) is much higher in small rural than in big urban communes (Strömberg, 1974b).

We also found that councillors differ relatively little from one another in the priorities they give to various public services. Neither in this case can the differences be traced to socioeconomic representativeness. When the councils' aggregate service priorities are compared with those of various social groups within the electorate, only small differences appear. This indicates that the councils, in spite of the biases in socioeconomic composition, do represent a broad spectrum of opinions in society. Even in this respect the party system plays a role, in that differences among the voters' service priorities are reflected by their representatives on the councils. On the whole, however, differences in service priorities are small—both with respect to different groups within the electorate and with respect to councillors vis-à-vis voters.

There are, however, important variations among different commune types with respect to service priorities—both at the council and the electorate levels. The service priorities observed in rural communes are quite different from those observed in urban communes. Thus, it appears that local recruitment of political decision makers and the links between local and national party systems ensure that important opinions in the electorate are reflected on the councils. This holds true in spite of the facts that the parties do not campaign actively in local areas, that voters are, by and large, not well informed about the local party system, that voters are generally not active in local political parties and organizations, and that councillors differ socially, economically, and educationally from those who elect them.

NOTES

1. This definition of local self-government was formulated on the basis of a study of Swedish public discussion on self-government and forms the basis of the general conceptual framework used by the Swedish Local Government Research Group (Westerstahl, 1967, 1970a, 1970b).

2. The town meeting system which prevailed in certain Swedish communes after the introduction of universal and equal franchise in 1919 and up until 1954 closely approached the ideal form of direct democracy. However, attendance at town meetings was low, which was one of the many reasons why councils were successively introduced.

In representative democracy, the circle of decision makers can be broadened by combining the local council with indirectly elected committees and boards possessing independent decision-making functions. But as a result of a massive reduction in the number of communes and communal representatives ensuing from two apportionment reforms, the possibilities for this kind of direct participation were radically reduced.

The number of communes was reduced by the first reform from 2,500 in 1951 to 1,000 in 1952 and to 278 by the second reform from 1965 to 1975. At the same time the number of representatives in councils, boards, and communes fell from 250,000 to 100,000 out of a population of eight million (Strömberg, forthcoming).

3. This general definition of representative democracy coincides with those prevalent in the literature. All more highly defined conceptualizations of representative democracy are susceptible to the influence of their own age and culture and also, e.g., to the structure of the electoral and party system actually in force. For a general review of the representation concept, see Pitkin (1967) and Birch (1971).

This view of democracy coincides with the "functionalist competition model," developed mainly by electoral researchers but also by other political scientists during recent decades (Pateman, 1970:3-13; Lewin, 1970:19 ff.).

4. The sample was constructed to permit a study of various systemic variables in order to predict the consequences of two large reapportionment reforms (cf. note 2). Size and density was supposed to be strategic since the number of inhabitants in the merged communes should rise at the same time as they should be composed of both urban and rural areas. To control for social-democratic majority was important since the political structure of the communes is strongly correlated with both size and density (Westerstahl, 1970a).

5. For a discussion of the right-left dimension in Swedish politics, see Särlvik (1969, 1974). The direct connection between party preference and attitude toward public sector expansion is studied in Christofferssen et al. (1972:71-94, 92-98).

6. A simple comprehensive index of this kind, calculated only as a mean, has shortcomings. The same index value can describe highly discrepant distributions. If half of any given group of individuals are in favor of raising local income taxes and the other half of reducing them, the group's index value will be the same as if they had all been in favor of keeping the tax at its present level. In spite of these shortcomings the index does yield a clear picture of crucial differences in attitudes between councillors and the electorate in various communes and parties, and the internal nonresponse rate in this question is low.

7. This list is open to criticism on the same grounds as many community power studies which have concentrated only on a few fields of activity. Admittedly, the problem here is different: not how to measure power exerted over communal activity, but the priorities given to various service fields. Yet the descriptive problem is similar. Thus, no fewer than four different fields of activity within social care have been selected, while schools, which in all communes constitute the largest share of the budget, have only one question devoted to them. Even if the choice of service fields cannot be regarded as strictly representative, an effort has been made to cover the whole realm of local government activities. Obviously, the

results refer only to these 12 fields, not to all communal activities. The crucial point, however, is that it was possible to study the same 12 service fields in all communes.

8. It should be noted that these questions measure the desire for changes in the *extent* of communal service. They do not measure attitudes toward the *form* taken by the activity. An individual, for instance, may be satisfied with the amount of housing being built within his commune, but still be critical of its being concentrated on multifamily houses. Such distinctions are not taken into account in the way the questions were constructed.

9. The procedure used here can be challenged from a methodological point of view. In order to obtain preference orderings, it is usual to ask interviewees to rank-order a number of items or, since this can often be difficult, to "force" a preference ordering by means of some special technique. One common technique is to have the interviewees compare items pairwise. Neither of these techniques was felt to be suitable, since their aim is to bring about rank-orderings for individual preferences. This often leads to the problem of individuals whose preference orderings are inconsistent or incomplete. Further, we here had to estimate the rank-ordering of public service fields for a group of individuals—e.g., voters in a certain commune. In these circumstances one must count on certain voters lacking any articulated attitudes to one or more service field. The technique used here, which can be said to resemble a plebiscite for each issue area, seems more realistic than a technique in which preference orderings are "forced." For a review and discussion of various preference-ordering techniques see, e.g., Kerlinger (1969).

An alternative technique for establishing preference orderings of communal value allocation issues with the aid of wholly open-ended questions—and of calculating the agreement between leaders and voters as an individual measure—is described by Verba and Nie (1972:302-304).

10. The Spearman coefficient was chosen because it is parameter-free and simple to calculate. Virtually all correlations given from here on are positive and significantly deviate from 0. For a discussion of various rank correlation measures, see Siegel (1956).

11. For a discussion of definitions and measures of social class, especially in Sweden, see Carlsson (1958). The ability of the occupational and social class classification worked out by the National Central Bureau of Statistics to explain political behavior is illustrated in a number of analyses of voting in the Lower Chamber of the Swedish Parliament (Särlvik, 1969).

12. Where dichotomies are concerned, we can choose between a number of different measures, all based on differences in percentages, the so-called index of likeness (Anderson et al., 1966:14). The representativeness index used here was chosen mainly because it can be used throughout to assess a specific group's under- or overrepresentation, irrespective of whether such an electorate group is defined by a dichotomy or an ordinal or nominal scale, and because the measure is easily calculated and interpreted. However, it cannot be used, for example, to summarize differences between two distributions. For a discussion of various measures of agreement see Alker and Russet (1964).

13. The construction of the Gini index is described by Alker (1965:36) and by other writers. But here Alker's calculation formula has been adjusted to allow for the fact that we have used a material that is divided up into classes. For a similar practical application of the Gini index, see Kjellberg and Offerdal, 1971.

14. The standardization in Table 10 is of course based on the assumption that the level of munificence in various income groups would not alter with a change in the income-wise composition of the councils. This assumption can naturally be challenged, especially for groups which at present are only weakly represented in the councils. The attitude of low-income councillors toward expenditure may, for example, be adapted to the attitudes of the overrepresented high-income councillors, and it may also be the case that it would change if a big increase occurred in the low-income group's representation in councils. The same line of reasoning could be applied to other weakly represented groups, e.g., women.

REFERENCES

ALKER, H.R., Jr. (1965). Mathematics and politics. New York.
ALKER, H.R., Jr., and RUSSETT, B.M. (1964). "On measuring inequality." Behavioral Science, 9(3):207-218.
ANDERSON, L.F., WATTS, M.W., Jr., and WILCOX, A.R. (1966). A legislative roll-call analysis. Evanston, Ill.
BARKFELDT, B., BRANDSTROM, D., and ZANDERIN, L. (1971). Partierna nominerar. Den kommunala Självstyrelsen 3. Uppsala.
BIRCH, A.H. (1971). Representation. London.
BIRGERSSON, B.O., FORSELL, H., ODMARK, T., STROMBERG, L., and ORTENDAHL, C. (1971). Medborgarna informeras. Den kommunala självstyrelsen 2. Stockholm.
CARLSSON, G. (1958). Social mobility and class structure. Lund.
CHRISTOFFERSSON, U., MOLIN, B., MANSSON, L., and STROMBERG, L. (1972). Byrakrati och politik. Gothenburg.
KERLINGER, F.N. (1969). Foundations of behavioral research: Educational and psychological enquiry. London.
KEY, V.O. (1961). Public opinion and American democracy. New York.
KJELLBERG, F., and OFFERDAL, A. (1971). "Politisk rekruttering—Nominasjoner ved kommunevalg." Tidskrift for Samfunnsforskning, 12(4):299-323.
LEWIN, L. (1970). Folket och eliterna: En studie i modern demokratisk teori. Stockholm.
LIPSET, S.M. (1959). Political man: The social bases of politics. New York.
MANSSON, L. (1970). "Rikspolitiska och lokalpolitiska attityder." Statsvetenskaplig Tidskrift, 73(1):1-34.
PATEMAN, C. (1970). Participation and democratic theory. Cambridge.
PITKIN, H.F. (1967). The concept of representation. Berkeley and Los Angeles.
SIEGEL, S. (1956). Nonparametric statistics for the behavioral sciences. New York.
STROMBERG, L. (1974a). "Väljare och valda: En studie av den representativa demokratin i kommunerna. Unpublished manuscript, Institute of Political Science, University of Gothenburg.
——— (1974b). "Participation and community structure." Pp. 80-95 in F.C. Bruhns, F. Cazzola, and I. Wiatr (eds.), Local politics, development, and participation. Pittsburgh: University Center for International Studies, University of Pittsburgh.
——— (forthcoming). "Local government reforms in Sweden."
SARLVIK, B. (1969). "Socioeconomic determinants of voting behavior in the Swedish electorate." Comparative Studies, 2(1):99-135.
——— (1974). "Sweden: The social bases of the parties in a development perspective." Pp. 371-434 in R. Rose (ed.), Electoral behavior: A comparative handbook. New York.
VERBA, S., and NIE, N. (1972). Participation in America. Political democracy and social equality. New York.
WESTERSTAHL, J. (1967). "Swedish local government research." Scandinavian Political Studies, 2:276-280.
——— (1970a). Ett forskningsprogram. Den kommunala självstyrelsen 1. Uppsala.
——— (1970b). "The communal research program in Sweden." New Atlantis, 2:124-132.

THE AUTHORS

ROGER W. BENJAMIN, Associate Professor of Political Science at the University of Minnesota, has worked on theoretical, methodological, and data analysis problems in comparative (cross-national) research. His book *Patterns of Political Development* (co-authored) and a number of articles in the *American Political Science Review, Journal of Conflict Resolution, Midwest Journal of Political Science, Comparative Political Studies,* and *Law and Society Review* reflect these interests.

BENGT OWE BIRGERSSON is Associate Professor of Political Science at the University of Stockholm. Since 1972, he has served as the Deputy Secretary for the Standing Committee on the Constitution in the Swedish Parliament. Between 1966 and 1972, he was engaged in research on local government in Sweden in connection with the Local Government Research Group at the University of Gothenburg.

FRANCES PENNELL BISH has done graduate work at the University of Washington and the University of Southern California in urban studies and public administration. Since 1974, she has been working with the Workshop in Political Theory and Policy Analysis at Indiana University on a major study of police services in metropolitan areas. Her major research interests include institutional analysis and design, urban problems, and civil and criminal justice administration and theory.

DANIEL J. ELAZAR is Professor of Political Science and Director of the Center for the Study of Federalism at Temple University. He also holds an appointment as Professor of Political Studies and Head of the Institute of Local Government at Bar-Ilan University, Israel. His major teaching and research interests include federal-state-local relations, state and local politics and urban and metropolitan problems in the United States, and problems of government, politics and culture in Israel. A contributor to over 70 books and scholarly journals, Elazar is also author or editor of 16 books including *The American Partnership: Federal-State Relations in the Nineteenth Century; American Federalism: A View from the States; The American System: A New View of Government in the United States;* and *Cities of the Prairie.* He currently serves as the editor of *Publius, The Journal of Federalism.*

BRUNO S. FREY received the degree of Lic. Rer. Pol. in 1964 and of Dr. Rer. Pol. in 1965 at the University of Basel. He is university professor at the University of Konstanz (West Germany) and managing editor of *Kyklos,* an international journal for social sciences. His main areas of research are economic theory and public finance. His publications include about 50 contributions to various scholarly journals in German and English on problems of economic growth, the economic theory of politics, the economics of environment, and the economics of peace research, and books on environmental economics, the economics of income distribution, and political economy.

ARTHUR B. GUNLICKS, Associate Professor in Political Science at the University of Richmond (Virginia), spent the 1975-1976 academic year as a Fulbright scholar in Göttingen. There he completed several articles on local government reforms and government and politics in West Germany. Earlier articles on government and politics in West Germany by the author have appeared in *Archiv für Kommunalwissenschaften, Journal of Politics, Comparative Politics,* and *Political Opposition and Dissent* (edited by Barbara N. McLennan). Gunlicks has also written articles on local government reform and campaign finance in Virginia. Currently he is working on a book on the subject of local governmental reforms in West Germany.

KENNETH HANF received his Ph.D. in Political Science and Public Administration from the University of California, Berkeley. Since 1973, he has been a Research Associate at the International Institute of Management in Berlin. During this time, he has conducted research on administrative development in advanced socialist systems with a particular focus on intergovernmental relations.

VINCENT OSTROM is Professor of Political Science and Co-Director of the Workshop in Political Theory and Policy Analysis at Indiana University. His principal works include *The Political Theory of a Compound Republic; The Intellectual Crisis in American Public Administration;* and *Understanding Urban Government: Metropolitan Reform Reconsidered* (coauthored with Robert L. Bish).

WERNER W. POMMEREHNE received his B.A. at the University of Basel in 1970 and his Ph.D. at the University of Konstanz (West Germany) in 1975. He is university lecturer in the Department of Economics of the University of Konstanz, teaching courses in public finance, the economic theory of politics, and local government. His research is in the fields of public goods, income distribution, and voter behavior.

PHILIP SABETTI is Assistant Professor of Political Science at McGill University, Montreal. His work is concerned with political theory and its application to the analysis of public service systems.

LARS STROMBERG received his Ph.D. from the University of Gothenburg in 1974. He is a Professor of Public Administration at the Aalborg University Center in Denmark. He was a research fellow with the Local Government Research Group at the University of Gothenburg and has also held various teaching and administrative positions with the Institute of Political Science at the University of Gothenburg.

RUUD J. VADER received his doctorate from the Vrije Universiteit in Amsterdam. His thesis was on the Soviet perception of West European integration, and his areas of interest include theoretical international relations and public administration. He has been an Associate Instructor of Political Science at Indiana University and a summer-associate with the Russian and East European Center at the University of Illinois at Urbana-Champaign. Currently, he is a research fellow for his alma mater at the Hoover Institution on War, Revolution, and Peace at Stanford University.